T0235402

The Nature of Classification

Also by Malte C. Ebach

BIOGEOGRAPHY IN A CHANGING WORLD (*with R. Tangney*, 2006)

COMPARATIVE BIOGEOGRAPHY: Discovering and Classifying Biogeographical Patterns of a Dynamic Earth (*with L. R. Parenti*, 2009)

FOUNDATIONS OF SYSTEMATICS AND BIOGEOGRAPHY (*with D. M. Williams*, 2008)

Also by John S. Wilkins

DEFINING SPECIES: A Sourcebook from Antiquity to Today (2009)

SPECIES: A History of the Idea, Species and Systematics (2009)

INTELLIGENT DESIGN AND RELIGION AS A NATURAL PHENOMENON (2010)

The Nature of Classification

Relationships and Kinds in the Natural Sciences

John S. Wilkins
University of Melbourne, Australia

and

Malte C. Ebach
University of New South Wales, Australia

© John S. Wilkins and Malte C. Ebach 2014

Softcover reprint of the hardcover 1st edition 2014 978-0-230-34792-2

All rights reserved. No reproduction, copy or transmission of this publication may be made without written permission.

No portion of this publication may be reproduced, copied or transmitted save with written permission or in accordance with the provisions of the Copyright, Designs and Patents Act 1988, or under the terms of any licence permitting limited copying issued by the Copyright Licensing Agency, Saffron House, 6–10 Kirby Street, London EC1N 8TS.

Any person who does any unauthorized act in relation to this publication may be liable to criminal prosecution and civil claims for damages.

The authors have asserted their rights to be identified as the authors of this work in accordance with the Copyright, Designs and Patents Act 1988.

First published 2014 by
PALGRAVE MACMILLAN

Palgrave Macmillan in the UK is an imprint of Macmillan Publishers Limited, registered in England, company number 785998, of Houndmills, Basingstoke, Hampshire RG21 6XS.

Palgrave Macmillan in the US is a division of St Martin's Press LLC, 175 Fifth Avenue, New York, NY 10010.

Palgrave Macmillan is the global academic imprint of the above companies and has companies and representatives throughout the world.

Palgrave® and Macmillan® are registered trademarks in the United States, the United Kingdom, Europe and other countries

ISBN 978-1-349-34515-1   ISBN 978-1-137-31812-1 (eBook)
DOI 10.1057/9781137318121

This book is printed on paper suitable for recycling and made from fully managed and sustained forest sources. Logging, pulping and manufacturing processes are expected to conform to the environmental regulations of the country of origin.

A catalogue record for this book is available from the British Library.

A catalog record for this book is available from the Library of Congress.

*To Gary Nelson, in whose trail-blazing footsteps
we dimly follow*

Contents

List of Figures

List of Tables

Acknowledgments

We are grateful for the discussions and information provided by other specialists both in the sciences discussed and the philosophy and history of those sciences. They are not responsible for any of the mistakes and provocations contained within. In particular we are most grateful to the following for discussions and assistance: Reed Cartwright, Steven French, Chris Glen, Jim Griesemer, Paul Griffiths, John Harshman, Mark Olsen, Brent Mishler, Dom Murphy, Gary Nelson, Eric Scerri, Emanuelle Serelli, Clem Stanyon, Tegan Vanderlaan, Kipling Will, and David Williams, as well as an anonymous reviewer. Thanks also to Christine Tursky for purchasing the first copy of this book a year in advance.

Introduction

A true classification includes in each class, those objects which have more characteristics in common with one another, than any of them have in common with any objects excluded from the class. [Herbert Spencer, *The Classification of the Sciences*][1]

Whereas in the thing, there is but one single unity, that sheweth (as it were in a glasse, at severall positions) those various faces in our understanding. In a word; all these words are but artificial-ltermes, not reall things: And the not right understanding them, is the dangerousest rocke that Schooles suffer shipwracke against. [Kenelme Digby, "Observations Upon Religio Medici", 1642][2]

When two species of objects have always been observed to be conjoined together, I can infer, by custom, the existence of one wherever I see the existence of the other; and this I call an argument from experience. But how this argument can have place, where the objects, as in the present case, are single, individual, without parallel, or specific resemblance, may be difficult to explain. [David Hume, *Dialogues Concerning Natural Religion*, 1779][3]

Classification has been missing from the philosophy of science for some time now. Oh, there has been a lot of discussion about natural kinds, and even more about the meaning and denotation of terms, both of which are a kind of classification, but there is little to no discussion of the process of simply classifying things, outside of psychological and cognitive studies. What scientists actually do when classifying natural objects in their domain, particularly in the absence of a working theory, is largely left untouched.

A century and a half ago, however, classification was a hot topic in the philosophy of science. It was seen as a major activity in science by scientists and philosophers alike. And it remained a topic of intense debate in a number of fields, particularly in biology. When philosophers discussed classification, though, they did so in the light of theories, and a good many scientific classifiers took them at their word, treating all classifications as being hypotheses or based upon theories.

In this book we will attempt to rediscover the role that classification plays in science. Consequently, what our target is, is natural classification, and not the classification of concepts or theoretical objects. It is what botanists, zoologists, and geologists did before they had explanations of phenomena. It is what psychology did before it had etiologies of pathological conditions. It is an underappreciated aspect of understanding how sciences develop.

We will first consider the relation of theoretical ontologies and empirical observations. Since most philosophy of science has treated ontologies in science as being theory-derived, and subsequent classifications based upon that, we thought it worthwhile to open some space for nontheoretical classification. In short, we argue that observation can lead to classification in the absence of a theory of a given domain. Once there are classifications in such theory-free fields, then explanations can be developed. Such classifications are phenomena to be explained.

In Chapter 1, we describe a way to conceptualize science as a field of possibilities from active conceptualization (theorization) to passive conceptualization (classification), and from active observation (experiment) to passive observation (pattern recognition of phenomena), setting up the scene for later chapters.

In Chapter 2, we introduce the notion of a natural classification and the role classification plays in sciences. We consider the difference between taxonomy and systematics, and introduce the question of theory-dependence of observation. The philosophical background of classification is introduced, along with the question of essentialism and natural kinds, which we replace in classification with the method of types.

In Chapter 3, we consider the sociological and phenomenal aspects of classification. The tribalism of taxonomy and systematics is discussed, leading to the tasks of classification, to order taxa and objects so that inferences can be made from them. Classing and ordering objects are distinct actions. We consider the iconographical representations of classification, and deflate "tree-thinking" somewhat. We note the influence on the thinking of classifiers of the ontological fallacy, believing that because we have given a name to a group we think we see, that that group must exist. Finally, we discuss names and nomenclature.

In Chapter 4, we consider what homologies and analogies are, in biology and other contexts. A homology is a relation from one set of objects or parts to another, a relation of identity no matter what differences of appearance or function exist in the parts or objects. Similarity relations are arbitrary, while homological relations are not. Homological relations are inductively projectible, based on consensuses of topographical agreement over time (causes) and space (forms). In biology, phylogenetic classifications are partial solutions to Goodman's grue paradox, since homologies are the right dependence relations to make inferences from. This is why, although the evolution of species is somewhat grue-some, we can make ampliative inferences about organisms. We finish with a discussion of Sober's "modus Darwin" being based on convergences (analogies) rather than the homologies (affinities) Charles Darwin actually employed.

In Chapter 5, we consider monstrous classifications, or misclassifications, which rely more upon facts about the observers and their predilections than upon the facts about the objects classified. Trashcan categories are common in science, but are aphyletic, in biological terms. We consider what is a natural classification, concluding that it is one based on a single cut of a classificatory hierarchy (monophyly in biology) rather than a mixture of artificial and natural characters. Natural kind classifications are grades based on the analogous characters preferred by a Theory.

In Chapter 6, we consider how abandoning the full theory-dependence of observation thesis (TDOT) affects our view of classification. We define a scientific Theory (capital-T) as something distinct from the notion that phenomena are observed based on prior criteria of salience to an observer, and adopt the Bogen–Woodward notion of a phenomenon as a pattern in data. A phenomenon, including a classification, is the explicandum that Theory explains. We then consider the question whether Theory from outside a domain of investigation counts as Theory Dependence within the domain, and thus ask what a domain is in science. We set up a "domain conundrum" – how can a science get started when there is no Theory of its domain?

In Chapter 7, we define a neutral terminology for classification across all scholarly fields. We have chosen to call it Radistics, from the Greek root word for "branch": radix (ῥάδιξ), to produce a schema into which the debates can be placed.

In Chapter 8, we note that what classifications contribute to the inferential process in science is that they allow us to locate the mass of data points observed without Theory in a broader pattern, and they guide Theory-building. Classification is not, in and of itself, Theory-building; nor is it free of Theory when Theory is available. However, if we have no Theory, or the Theory is contested, then we should recognize that a classification scheme is a statement of what we do know, and rest easy in our ignorance of what we do not.

This is a philosophy book, written for philosophers and scientists alike. It is not a book for scientists to appeal to in justification of their science. To paraphrase Feynman's famously attributed saying,[4] philosophy is about as useful to science as ornithology is to birds. Science is in the business of finding out about the world and explaining it. Philosophy is in the business of taking concepts, and the words used to express them, and trying to make sense of them, to stress-test them. Locke famously referred to philosophy as clearing the ground of science,[5] and as Wittgenstein wrote

> In philosophy we are not, like the scientist, building a house. Nor are we even laying the foundations of a house. We are merely "tidying up a room".[6]

While some philosophers also do science, and many scientists also do philosophy, these are distinct undertakings. The ornithologists do not help the birds fly. Sometimes, when the birds are reflecting upon themselves, ornithologists can help. And the birds, in this case, are self-reflective. More than most philosophical issues, classification is one that scientists argue over, attend to and extensively develop. If scientists themselves do this philosophy, it is only polite for philosophers to return the favor. This, then, is philosophy for the birds.

Philosophy of science has unduly neglected classification in the natural sciences for over a century, excepting in debates over biological taxonomy. Outside of biology, and even within it at times, classification has almost uniformly been treated as a matter of psychology, convention or of sociology. In the literature of the philosophy of science, natural classification has been considered to be theory-driven, a matter of determining the natural kind ontology of a domain or discipline. In general metaphysics it has been a question of universals, terms which cover many individuals.[7] Otherwise, classification was considered to be a kind of pre-scientific holdover, or a heuristic stand-in for what actually was proper scientific method; it was a naive inductivism that could be dispensed with, something that we left behind in the nineteenth century.

The aim of this book is to revive an interest in the philosophy of natural classification, and to locate it well and truly within the scientific process, both as a scientific activity that has payoffs for theory without necessarily being derived from it, and as a heuristic for discovery of phenomena that themselves call for theoretical explanation. We are not inclined to give technical details or justifications for this or that methodology – very often these are still being tested and assessed by the practitioners of the sciences – and they are grist to the mills of specialists in statistics and mathematics, as well as the logic of probability and inference. However, the role of classification in science itself, by whatever techniques, needs to be restored to the attention of general philosophy of science, and not merely as something that only biologists and medical researchers do.

Infamously, and apocryphally, Ernest Rutherford would tell his students, "All science is either physics or stamp collecting!" This view both represented and encouraged the view that "stamp collecting" was a pursuit for lesser scientists and sciences, and stamp collecting was taken by most people to mean the contingent, particularistic, and merely taxonomic sciences. Since by that time, around the early years of the twentieth century, classification had become the domain of the librarian, the analytic tradition – with one notable exception, John Henry Woodger – ignored even biological classification, and even biologists came to believe that the identification of taxonomic groups, from species up, was a matter of convention.

But this cannot be true. Not only have biologists continuously and more or less consistently classified roughly the same sorts of things, species, for the duration of the modern biological era, but other sciences, including physics, have done it as well, and this has been enormously fruitful, driving both theory and experiment. Classification is a persistent moment in a science's development. However, it does not occur in the same manner in every theory's or discipline's development. Why this is the case will be a target of the discussion that follows.

Because the focus has, for the last century, been so restrictedly upon language and psychologism, the notion of a "natural" classification has very largely gone by the wayside in the philosophy of science. When it has been considered at all, it is in terms of natural kinds, with theoretical or other essences, as in the work of Brian Ellis and Joseph LaPorte.[8] This is not a book about natural kinds.

At the turn of the nineteenth century, though, classification was a focal issue debated at length by a great many scientists, many of them mineralogists, chemists and physicists, as well as biologists. Philosophers like William Whewell in 1840,[9] through to W. Stanley Jevons nearly forty years later,[10] wrote about the natural classification of things. By the turn of the twentieth century, however, the most widely influential text on classification was a librarian's,[11] and the consensus thereafter was that the units of classification in biology, species, were merely conventionally defined objects used, as Locke said, only for communication, a view repeated by the great evolutionary biologist John Maynard Smith nearly sixty years afterwards.[12]

Interestingly, where classification did play a large role in philosophy in the latter half of the nineteenth century was in the classification of the sciences themselves. From Auguste Comte, to Herbert Spencer, to Jevons, a major topic was the classification of the subject matter, method and domain of the sciences and their relation to each other. This explains why classification in philosophy ended up becoming classification of books and topics; it was seen as not so much a matter of nature, but of philosophical construction.

A further problem is the widespread belief that classification is solely a matter of convention. Even in biological systematics, group terms like species are often regarded not as natural objects or classes or kinds, but as constructs of sheer convenience. This is extended to other sciences when there is no theoretical kind or essence involved. In general, definitionalism has been largely abandoned in the social sciences, where it is called "essentialism"; in psychology, where it is called the Classical Theory; and in philosophy, where it is often thought to have been defeated by Bertrand Russell, Ludwig Wittgenstein and Hilary Putnam.[13] In its place are such theories as stereotype theory, prototype theory, and exemplar theory.[14] These are theories of meaning and intentionality, and have only a tangential connection with classification in the natural sciences, being issues in the philosophy of language and psychology. As such, they are scientific classification theories only when the scientific question is the development and evolution of language and mind. It is crucial not to take a psychologistic approach and project properties of cognitive dispositions onto scientific, that is to say, natural, classification.

Our aim here is to raise and consider some of the old questions of classification in the modern context, to get at the nature of classification in the natural sciences. We think that this underplayed aspect of science makes sense of theoretical and historical development of the sciences, and of the modes of discovery that are largely mysterious in its absence. Moreover, natural classification is a tool of inference and testing, and not, as the common view has it, a hypothesis of either history or process; it allows us to test hypotheses about these matters.

The authors are a working scientist (Ebach) and a philosopher and historian of biology (Wilkins) and so the structure of the argument is in some ways a dialogue between them as surrogates for those working in the sciences, at the coalface, as it were, attempting to generalize theoretical issues, and with those who approach the sciences from a meta-level perspective. We do not entirely agree on every topic or argument raised here, and we will attempt to make our differences known and the reasons for them clear in the footnotes. Consider this book as an invitation to engage in a discussion rather than a supposedly authoritative last word. If, for philosophers, it is written in an overly nontechnical fashion, be assured that for scientists it will be seen as too technical, and so we hope to satisfy nobody.

Moreover, this is not a work of general abstract philosophy of language, epistemology or metaphysics. We consider these matters only so far as they are relevant to understanding how natural classification plays a part in science. It is Wilkins' opinion that such matters are largely decoupled from the broader concerns of epistemology and metaphysics, and can be bracketed out in this context, except where noted in the text.

The Earth occupies an almost infinitesimal amount of the observable universe, for less than a third of the time the universe has existed, and yet almost all of that which science classifies is found only on the Earth (hence the icon on the cover). One might even think that natural classification is in fact just Earth science. We argue that while the general sciences tend to classify by analogy, the palaetiological sciences, the historical causal sciences, tend towards homological classification. Case studies are scattered through the book as boxes. They include the periodic table (initially analogical), clouds, soils, and of course biological taxa. We offer a short discussion regarding classification in psychology, and in particular psychological diseases in the *Diagnostic and Statistical Manual*, or *DSM*. We might have considered many other similar topics: geology, biogeographic regions, minerals, chemical compounds, biochemical structures, climatological regions, and physiological diseases to name only a few. But we have chosen these cases for now. We hope that other researchers will be inspired to follow this line of research in these and many other fields involving natural classification.

Notes

1. Spencer 1864.
2. We owe this quote to Jeb McLeish.
3. Part 2, italics original. Page 65 in the 1779 second edition (Hume 1779).
4. Most likely due to Steven Weinberg, although it may have been coined in a review by McHenry 2000 summarizing Weinberg. In any case it is a reuse of a much older saying about aesthetics and artists by Barnett Newman:"I feel that even if aesthetics is established as a science, it doesn't affect me as an artist. I've done quite a bit of work in ornithology; I have never met an ornithologist who ever thought that ornithology was for the birds." He would later hone this remark into the famous quip, "Aesthetics is for the artist as ornithology is for the birds". See <http://www.barnettnewman.org/chronology.php>.
5. *Essay Concerning Human Understanding*, Introduction: Philosophy is "employed as an under-labourer in clearing the ground a little, and removing some of the rubbish that lies in the way to knowledge".
6. Monk 1990, 298f.
7. See Armstrong 1978; Sloan 1985.
8. Ellis 2001, 2002; LaPorte 1996; LaPorte 1997, 2004.
9. Whewell 1840.
10. Jevons 1878.
11. Richardson 1901.
12. Maynard Smith 1958.
13. Wilkins 2013.
14. Cf. Prinz 2002.

1
The Nature of Science

The dance floor of science

About thirty years ago there was much talk that geologists ought only to observe and not to theorize; and I well remember someone saying that at this rate a man might as well go into a gravel pit and count the pebbles and describe their colours. How odd it is that anyone should not see that all observations must be for or some view if it is to be of any service. [Charles Darwin[1]]

... the work of theory and observation must go hand in hand, and ought to be carried on at the same time, more especially if the matter is very complicated, for there the clue of theory is necessary to direct the observer. Though a man may begin to observe without any hypothesis, he cannot continue long without seeing some general conclusion ... he is led also to the very experiments and observations that are of the greatest importance ... [and] the *criteria* that naturally present themselves for the trial of every hypothesis. [John Playfair[2]]

In this chapter we describe a way to conceptualize science as a field of possibilities from active conceptualization (theorization) to passive conceptualization (classification), and from active observation (experiment) to passive observation (pattern recognition of phenomena), setting up the scene for later chapters.

According to traditional philosophy of science, by which of course we mean what Wilkins was taught as an undergraduate,[3] what science does is to develop, test, and argue over theories. Oddly, what a scientific theory consists of is rarely discussed, although there is a consensus that a theory is a formal model of a family of models with ancillary hypotheses and interpretations of some kind.[4] In this book we shall consider "theory" to cover any abstract representation or part of such an abstract representation including models. However, the focus has been on theories at least since John Stuart Mill's *A System of Logic* in 1843,[5] especially once that work was adopted as the basis for the burgeoning analytic philosophy movement in Britain and America, and the subsequent development of logical positivism and its heirs and successors.

Positivism was a two-dimensional or linear historical progressivist view about science. Comte himself held that societies moved through the theological, the metaphysical and then the positive stages. Likewise, individual sciences were also held to develop this way. This progressivism persisted long after positivism died or transmuted into logical empiricism. Even as the Baconian idea of sciences developing from masses of naive observation into laws and theories was being abandoned, people still held that there was a constrained historical sequence for the development of sciences. For example, Thomas Kuhn's "normal science/revolutionary science" distinction, and in particular his "evolutionary metaphor":

> Imagine an evolutionary tree representing the development of the modern scientific specialities from their common origins in, say, primitive natural philosophy and the crafts. A line drawn up that tree, never doubling back, from the trunk to the tip of some branch would trace a succession of theories related by descent. Considering any two such theories, chosen from points not too near their origin, it should be easy to design a list of criteria that would enable an uncommitted observer to distinguish the earlier from the more recent theory time after time. Among the most useful would be: accuracy of prediction, particularly of quantitative prediction; the balance between esoteric and everyday subject matter; and the number of different problems solved. – Those lists are not yet the ones required, but I have no doubt they can be completed. If they can, then scientific development is, like biological, a unidirectional and irreversible process. Later scientific theories are better than earlier ones for solving puzzles in the often quite different environments to which they are applied. That is not a relativist's position, and it displays the sense in which I am a convinced believer in scientific progress.[6]

History moves forward. Unfortunately, this is not necessarily true of biological evolution, and there is no reason to think it is true of cultural evolution either, so why should it be true in science? Why must science follow a set trajectory? Why can scientific history be Whiggish when the rest of history cannot?[7] The presumption here is that the history of science is constrained to develop in particular ways. This is just false. Philosophy of science once assumed that the early stages of a scientific discipline are marked by basically wandering about observing stuff until a theory, hypothesis or law suggests itself. Then it gets tested. Let us suppose that constructing a theory is a phase of active conceptualization. We take our ideas and put them together into a coherent and explanatory structure, and having done so, we subsequently run experiments to test this and ensure that what we test is just the theory, and not confounding variables. This, very roughly, is Popperian falsification. And it is, most of the time, precisely not what most scientists do. Karl Popper and his followers were criticized for failing to deliver any vestige of a logic of discovery, despite the English title of Popper's masterwork. In fact, discovery was regarded as accidental, if anything. The real work was in the construction and testing of hypotheses, leading to models and thence to theories.

Philosophies of science tend to distinguish between the conceptual and empirical aspects of science. We might represent this as a field of possibilities, in which one axis is conceptual development, and the other of empirical observation. Even views based upon the theory-dependence of observation make the distinction, if only to assert the priority of one over another, so let us take this as a first approximation. Conceptual tasks are themselves divided into theoretical and classification tasks, the first being a representation of phenomena, and the second supposedly a systematization of the results of the dynamics captured by the theory/ model.

These two conceptual tasks are usually held in opposition, although again some subordinate the one to the other, mostly holding that theory determines the sorts of categories into which things get sorted. More rarely, holding that one's ontology, or classification of possible types of things, determines or constrains theories. Let us visualize each task as a set of goals connected by the common feature of being conceptual, like a dumb-bell. Empirical tasks, similarly, are divided into naive observation and more informed experimental testing, which involves knowledge of the theory. So, on this view of science, the "moments" between which scientific behavior "moves" look like Figure 1.1.

The Baconian Cycle (that is, the view held by those who thought they were doing Baconian induction) is shown in Figure 1.1 as sequence B, while the Popperian Cycle is shown as sequence P (Popper dismissed classification the way Rutherford did, as mere stamp collecting). Of course,

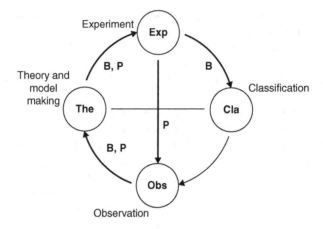

Figure 1.1   The Baconian (B) and Popperian (P) Cycles

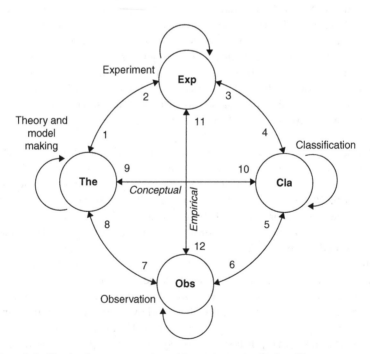

Figure 1.2   The twelve movements and four moments of scientific processes

all views of science hold it to be an iterative process, so if the results are not satisfactory, a movement can be indefinitely repeated. Let us now change the metaphor to a Cartesian graph with two axes: conceptual and empirical. A simple Popperian cycle, with theory-dependence, might inscribe a quite complex trajectory. If one is trying to work an experiment, or reformulate a model, loops will occur. The permutations for an extended process can become very extensive indeed. But what happens if we allow, as we surely must, that observation can inform classification, or that it can even, as Francis Bacon had it, inform a theory or a model? Let's fill in all the blanks (Figure 1.2).

This characterizes all possible movements for any autonomous scientific process, from within the mental processes of a researcher, to the work of a research group, to a general research program, to the activity of an entire discipline. The influences from one scientific task ("moment") to another are represented as outputs feeding into inputs ("movements") (Table 1.1). The processes within the moments include data analysis and other transformations local to that aspect of science. Experimental techniques, classification sequencing, observational processes and technologies, and theory-building are all aspects of their respective black boxes (shown above as white circles).

On the B and P cycle and traditional post-Popperian views, we have basically ignored half or more of what it is that science does! There are twelve possible pathways for the methodological influence of one task type to lead to results in another, plus the four pathways of self-correction and revision, by, for example, taking observations again to ensure precision and accuracy. It seems quite feasible to think both that observation might be influenced by theoretical assumptions and expectations, and that we might develop theory-free classifications on the basis of our experience and the classification systems, neural and analytic, that we apply to such data.

The combinatorial possibilities for any realistic sequence of research are immense, and if we add the possibility that these moves might occur in parallel, and that the moves might be distributed (research groups typically have many brains to do their work on, especially the more pliable brains of doctoral and postdoctoral students), we begin to have an extremely complex dynamic system. We leave it as an exercise to the reader to work out how complex.

It doesn't end there. Many recent treatments of science claim that the history of a science is not simply determined by its internal methodological workings (internalism) but that even the very facts with which it deals, no less than the ways experiments are run, theories are constructed,

Table 1.1 Moments of scientific activity

	Experiment	Classification	Observation	Theorizing
Experiment	Revision and correction	3. Experiment can suggest classification category	11. Experiment can restrict or guide "naive" observations	2. Experiment can restrict theoretical range, or disconfirm theory
Classification	4. Classification can suggest things to measure and expect	Revision and correction	5. Classification can guide the evidence sought	10. Classification can restrict or guide the ontology of a theory and the explanatory categories used
Observation	12. Naive observation can influence the data used in experiment	6. Classification can be based on naive observations made pretheoretically	Revision and correction	7. Theoretical predictions can fail to be borne out in observation
Theorizing	1. Theory can specify legitimate experimental protocols and approaches	9. Theoretical variables can become classification categories	8. Observation can depend upon the ontology and methodology of a theory	Revision and correction

and classifications developed, is a process influenced entirely or in part by its social and political milieu, for example in "actor-network theory".[8] Even if we are able to hold that presumption to a minimum, every discipline, research program, or laboratory is influenced at some or each point by external factors such as funding, engineering and technological resources and equipment, and other disciplines, and this must be accommodated in a realistic view of science.

The introduction of computers is a case in point; with effects of all three kinds – you have to pay for the hardware, the programmers, and bioinformaticians, and you are limited or freed according to the capabilities (and theoretical developments) outside your particular specialty. This view, if pushed to the limit, is externalism. Some versions of externalism, constructivisms, go so far as to make scientific observations and theories depend entirely upon political, religious and philosophical ideologies. Scientists are just apologists for a worldview and their data are "negotiated" by the community of similar believers. This view goes far beyond the "theory-dependence of observation" doctrine of Popper towards an "ideology-dependence of observation" stance. It often takes the form that evidence is selected in order to bolster the community's core doctrines, like finding evidence for natural selection is supposed to shore up capitalist free-market economics.

Data is not the basis of science for constructivists – instead the basis is the goals set by prior theoretical commitments based on some wholly external set of beliefs or conditions. We can reasonably reject the strong claims of naive constructivism, but there remains a large residual point that cannot be ignored. Social and cultural conditions do indeed influence scientists, theories, research programs, and institutions. Often this influence is exercised by biases in funding and support, but it is also, throughout the history of science, derived from ideas, especially of philosophy, and these influences can be creative and positive, as well as "unscientific". From Newton onwards, inspiration for hypotheses have been drawn from astrology, alchemy, theology, economics, literature and metaphysics. Sometimes attention is drawn to data that had not been previously examined, or were ignored as anomalies, because they serve political ends (as often happened under the Soviet Leninist regime). These influences can act as either positive and negative inputs into the scientific process, and we therefore must add to our schematic both these inputs, and the corresponding outputs – the influence of theoretical and taxonomic terms, observations and experiments – on the broader social and cultural systems in which it is situated (Figure 1.3).

From the perspective of the social milieu, science is an active and dynamic process that transforms data inputs and resources into

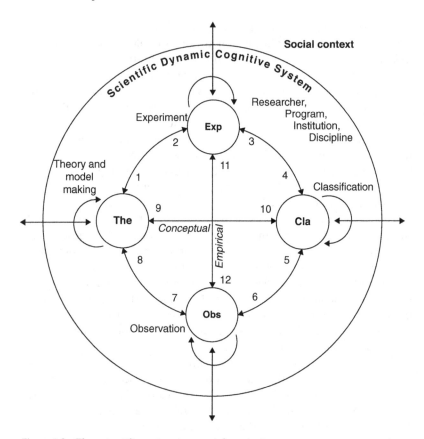

Figure 1.3   The scientific system in a social context

representational outputs and products. Science, at any level from researcher to discipline, is a dynamic cognitive system.[9] It is a formal analogue to any other cognitive system, such as a classifier or heuristic system implemented on a computer. It should be obvious that the scientific dynamic cognitive system also feeds back into society.

If we think of this field of possibilities as a dance floor, we can get a handle on how it is that scientific agents – individual scientists, laboratories, research programs, schools of thought, institutions and disciplines – behave. If there is no simple or singular trajectory around this dance floor, we might expect that individuals and all other agents in science, right up to disciplines, will tend to congregate closest to the action, but where there is a reasonable chance of success (the dance partners being, here, funding, students, collaborators, and so on). In other words, they will look for some elbow room, and rewards. If the choice of dance partners is so empty in one region that it won't pay to go to that corner, then the science will tend to congregate in the areas where there is a greater choice of partners. When the density of dancers to partners is so high that your chances of competitive success are low, you will move to a part of the floor where there is a better choice and chance.

Consequently, a new scientist, or discipline, or school of thought may embark on their science in a region not traditionally discussed by the philosophers, and one of those is the corner in which researchers do naive observation (passive observation in the field, usually), and classification. Or one might classify on the basis of experimental work, or make theory on the basis of the prior passive observation, and so on. Theory may lead to observation of either kind: "go out and see if guppies in Jamaica match the model of selective balance for mate attraction and predator avoidance", as ethologist John Endler did.[10] A novel field, however, lacks dance partners to speak of, and so naive classification may be all one can do, to begin with. Our physics-fascination has led us to think that all science is theory and experiment, when even in physics that is not true.[11]

Classification sets up problems that theory has to deal with. A classification is the explanandum, that which is to be explained; theory is the explanans. However, theories and classifications are also historical objects; they change over time, and so a theory at *t* can influence a classification at *t* + *n*, and the classification can then influence the theory at *t* + *n* + *m*, and so on. Science is an iterative process over time. We suggest that science is a field of possible moments, with no set trajectory over the "dance floor of science", but only an individual history of movements. Some might object to this, and argue that there is no real difference between classification and theory-building. We disagree, but first we had better make it clear that this is a continuum, and that any actual event in the history or development of science or a scientific career is almost never going to be purely theory or classification building. These are poles that are asymptotically approached on occasion.

One reason why the "theory-dependence" hypothesis was so successful is that it is almost trivial to find elements of theory in any set of observations, and likewise the inverse is true: it is almost trivial to find elements of classificatory activity in a theory. On our account, this is inevitable, not least because nobody ever starts a scientific process tabula rasa. We have prior conceptual commitments at any point, some being mediated by evolution itself (our *Umwelten*[12]), and some being mediated by the general cultural context in which we begin. For example, when biological classification began in the fifteenth century, it was built on the traditions of herbals and bestiaries that were the medieval heritage, but the recognition of kinds of plants and animals was not itself theoretical.

Even within science, in fact especially so, when we begin upon a new subject of investigation, we bring to that activity all the prior theories, methods, techniques, industries and educational substrates that exist at the start. Nobody commences knowing nothing. The theory-dependence view has it that we can only begin to learn about the world because of our theories. On our account, we learn because we are learning machines;[13] and prior knowledge always biases how we do that. However, we are not constrained in what we can observe merely by conceptual commitments, nor are we unable to "see" phenomena because theory doesn't permit it. Either of these would inhibit our ability to learn.

And learn we do. One way we do so is by identifying phenomena that are regular in some domain, through pattern matching. If there is no theory, or the theory is inadequate, these phenomena trigger the development of explanations. So how does classification, which is one kind of conceptual construction, differ from theory-building, which is another? Our answer is based on what the metrics on the axes are: each asymptote is either active or passive. Experiment is active intervention,[14] but field observation is often merely passive (and of course one can do field experiments). Theory construction, which is explanatory, is active conceptualization, but classification is more passive; it is "handed to you" by your cognitive dispositions and the data that you observe. Without adopting the observation sentences approach of logical positivism, it remains true that we do have a distinction between observation and explanation, even if nothing is ever purely the one or the other.

To classify is to order the data, to find regularities even if you have no direct idea why they cluster so. It is to find identity classes empirically, setting up some problem or thing in need of explanation. Darwin used the regularities of largely atheoretical classification, which he referred to in the *Origin* as "the subordination of group under group", as the explanandum that his "theory" of common descent by branching evolution explained.[15] The empirical regularities of the periodic table were subsequently explained, first, by valency theory, and then by quantum mechanics.[16] There is a distinction between observation and sorting on the one hand, and model building and explanation on the other.

We haven't specified what the "context" is into which we place this scientific dynamic system. We haven't done this because the context is an empirical matter in each case. There are no generalizations about this that are unexceptional. Sometimes a science will run more or less independently of its popular cultural context, and at other times a science will be beholden to its cultural context independently of the internal issues of the science. Consider, for example, the difference between late nineteenth century physics and psychoanalysis. The former was almost acultural, equally practicable by western, Indian, and Asian workers of all genders,[17] while the latter was tied to the cultural context, first of Viennese society, and later of New York and of middle-class America in general. A recent paper on WEIRD ("Western, Educated, Industrialized, Rich, and Democratic") psychological studies indicates that this remains an issue in psychology.[18] To think there is a general, universal and consistent cultural context for science is, we believe, a holdover of Comtean positivist thinking. You want to know what the relevant context was for the Hubble telescope, or for the discovery of aspirin? Go look. The schematic here merely indicates the general classes of relations of external and internal movements in the science.

The classification of clouds

Clouds were regarded as so subjective, fleeting and resistant to classification that they were a byword for the failure of empirical classification, until Luke Howard in 1802 proposed the foundation for our present system of cloud classification (in competition, although he did not know it, with others in Europe, and on the heels of Hooke and later meteorological language proposals, including one by Lamarck the same year).

Howard's proposal, like Lamarck's, was driven solely by empirical observations. No experiment was possible with clouds (although there were some schemes for building cloud producing machines early on), and there was no real theory as such, just a desire to, as Lamarck said, note that "clouds have certain general forms which are not at all dependent upon chance but on a state of affairs which it would be useful to recognise and determine".[19] In short, this is an example of a classification scheme without much if anything in the way of Theory.

Howard proposed seven classes (genera) of clouds – three "simple modifications", cirrus, cumulus, and stratus, two "intermediate modifications", cirro-cumulus, and cirro-stratus, and two "compound modifications", cumulo-stratus and cumulo-cirro-stratus, or nimbus. His criteria used apparent density, elevation, height, and whether it produced rain. Particular types of clouds were called, following the logical and Linnaean examples, "species". He also devised our present system of signs for these cloud types, and proposed a correlation with certain types of rain and clouds. Now meteorologists could communicate and seek explanations and presently the International Cloud Atlas is the global standard for identifying clouds.[20]

This is a classic example of an empirical passive classification. Although the hydrological cycle was of ancient vintage, the direct theory of clouds, such as it was, had to await the hypothesis of the thermal theory of cyclones and cloud formation.[21] Similar passive classifications were done for wind, resulting in the Beaufort Scale.

Howard's scheme outcompeted Lamarck's largely because of its technical terminology and signs. Lamarck's was too French and odd even for them. Howard's system gained great acceptance. Johann Wolfgang von Goethe, even wrote a poem in Howard's honor, as well as contributing "Towards a Study of Weather" in which he briefly discusses Howard's categories of clouds and a basic law of weather.[22]

Table 1.2 Cloud classification

"Family" or Étage (C_H)	Genera	Species	Coding	Varieties	Coding
High level (C_H)	Cirrus (Ci)	fibratus (fib) uncinus (unc) spissatus (spi) castellanus (cas) floccus (flo)	$C_H = 1 / 4$ $C_H = 4$ $C_H = 3$	intortus (in) radiatus (ra) vertebratus (ve) duplicatus (du)	→
	Cirro-cumulus (Cc)	stratiformis (str) lenticularis (len) castellanus (cas) floccus (flo)		undulatus (un) lacunosus (la)	→
	Cirro-stratus (Cs)	fibratus (fib) nebulosus (neb)	$C_H = 6$ or 7 $C_H = 7$	duplicatus (du) undulatus (un)	→
Medium level (C_M)	Altocumulus (Ac)	stratiformis (str) lenticularis (len) castellanus (cas) floccus (flo)	$C_M = 8$ or 9 $C_M = 8$ or 9	translucidus (tr) perlucidus (pe) opacus (op) duplicatus (du) undulates (un) radiatus (ra) lacunosus (la)	$C_M = 3$ or 4 $C_M = 5$ or 7
	Altostratus (As)	→	→	translucidus (tr) opacus (op) duplicatus (du) undulatus (un) radiatus (ra)	$C_M = 1$ $C_M = 2$
	Nimbostratus (Ns)	→	$C_M = 2$	→	→

Low level (C_L)				
Stratocumulus (Sc)	stratiformis (str) lenticularis (len) castellanus (cas)	$C_L = 5$ $C_L = 5$ $C_L = 5$	translucidus (tr) perlucidus (pe) opacus (op) duplicatus (du) undulatus (un) radiatus (ra) lacunosus (la)	→
Stratus (St)	nebulosus (neb) fractus (fra)	$C_L = 6$ $C_L = 7$	opacus (op) translucidus (tr) undulates (un)	→
Cumulus (Cu)	humilis (hum) mediocris (med) congestus (con) fractus (fra)	$C_L = 1$ $C_L = 2$ $C_L = 2$ $C_L = 7$	radiatus (ra)	→
Cumulonimbus (Cb)	calvus (cal) capillatus (cap)	$C_L = 3$ $C_L = 9$	→	→

Note: The hierarchy works from left to right (Family, Genus, Species, Variety). For example a High Level (3–8 km) transparent, whitish veil of hair-like or smooth appearance, either partially or totally covering the sky (Cirro-stratus), which form straight to curved filament that do not end in hooks (fibratus) may be coded as CH = 6 (CH = 0 being no clouds at that level). Cirro-stratus fibratus may also be written as Cs fib. and, like a biological taxonomy, will have a designated author (that is, Cirro-stratus fibratus Besson 1921). [a]Note that cloud families were abandoned in 1956 and replaced by Étage. The coding does not only apply to types of clouds, but to the sky as a whole over time as well as processes such as wind, and so on. Codings produce similar clouds may be due to different processes as in the case of Nimbostratus and Altostratus opacus (CM = 2), which may have arisen due to other cloud formations.

[a]Besson 1921.

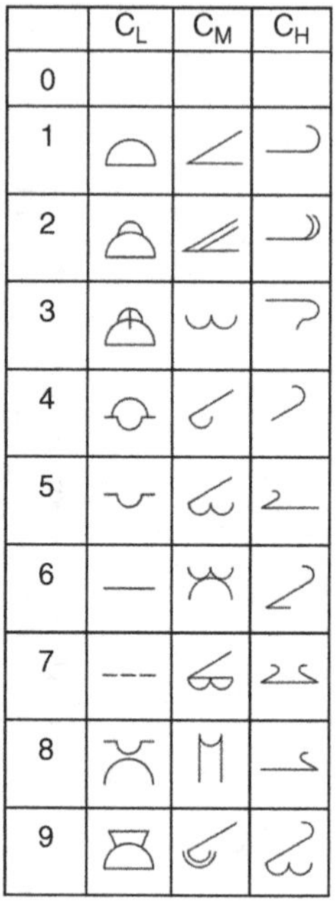

	C_L	C_M	C_H
0			
1			
2			
3			
4			
5			
6			
7			
8			
9			

Figure 1.4   Symbols for clouds corresponding to the different figures of the CL, CM and CH codes

Redrawn from the World Meteorological Organization 1956.

Classification and incommensurability

The existence of meaningful observation-based classes and the associated language outside the domain of theory suggests a partial solution to the problem of incommensurability. Kuhn had argued, as Ludwik Fleck and others before him, that a global theory has its own language, and if the intensional meaning of its terms is derived from the role they play in the formal theory itself, then there is no way to compare across a theory-change, as terms like "mass" simply are not commensurable.[23] The theory is self-supporting, and exhaustive: there are no ways that can be used to evaluate the theories with respect to each other, as meaning is solely derived within each "paradigm".[24] But if we have a relatively theory-free set of observationally derived classes, then we can compare how well each theory does at explaining them without the threat of vicious circularity or pure[25] relativity. If, as we argue, these observational classes form the explicanda of theory, then it doesn't pose a problem that the theoretical terms are internal to the paradigm, so long as some general criteria for satisfactory explanation exist along with these explicanda. Natural classification thus provides a bridge between theories, at least to the extent that it approaches the naive and passive observational asymptote. This is made most clear by the way Darwin treated the classifications of his day as explicanda for any theory of the origin of species and higher taxa. As the classification was produced on the basis of a mixture of convention and practical (that is, non-theoretical, in the main) natural history, he may have been inclined to dismiss it as artifice and begin again; but instead he adopted it as worthwhile, and not merely because he was part of that establishment.[26] He thought, as did most taxonomists of the day, that while the ranks of systematics were artificial, the objects so named were not, and the subordination of group under group a "grand fact" to be explained, as he explicitly says:

> Thus, the grand fact in natural history of the subordination of group under group, which, from its familiarity, does not always sufficiently strike us, is in my judgment fully explained.[27]

While not all inter-theoretic terms can be grounded in the usage of pre-theoretic terms of natural classification in the sciences, a great deal can be if classification can lead to theory rather than having to be derived from it. The paradigm approach is a kind of "worldview-ism", in which the conceptual *Weltanschauung* of the scientist bounds all they can say or see, not unlike the Sapir–Whorf hypothesis. However, nobody begins doing science, either historically or personally, from a position of complete ignorance. We have prior dispositions from our biology, our evolved psychology and cognition, as well as the relatively untheoretic shared folk knowledge of the society in which the scientist is situated. A straight *salva veritate* substitution of a term like "organism" or "mass" might not be possible just in terms of the theories themselves, but if they are part of a wider cognitive dynamic, including the rest of science outside the domain the theories apply to (as in optics providing a justification for Galilean cosmography), we can evaluate them without begging the question. When Wittgenstein wrote that if a lion could talk, we could not understand him, he overlooked the shared biological history we have with lions. The lion would be saying, "My mates! My food! My hunting grounds! My cubs!" and we'd understand this without trouble.[28] Likewise, a shared vocabulary and cognitive process apart from theoretical inference and terminology reduces the incommensurability. Natural classification has that role.

Observer systems and classifiers

It is a truism almost not worth noting that classification is an activity undertaken by classifier systems, usually specialists in a field. However, classifications are often treated as more or less successful representations of the objective state of the world, as if they were handed from on high by God, Theory, or some other divine agency. It is therefore worth making this truism explicit.

A classification is an abstract representation. It is a set of classes that are held to obtain in a domain, or universe of discourse.[29] We will consider later what a domain is, but we need some general notion of the actors involved in classification. We can speak, therefore, of classifiers and observer systems. A classifier is a system, which might reside between the ears of a trained taxonomist, or might be instantiated as a computer algorithm, or even as a social practice of which nobody is entirely conscious of. The speciality of, say, bryophytology, or the biological discipline that studies mosses, is as much a classifier, as a bryophytologist is, like Brent Mishler. Likewise, an observer system is a system that stands in a particular relation to the empirically accessible world for a domain. That relation is one of salience: an observer is disposed to see some data as relevant and significant, and to ignore other data. In both cases these systems can exist at various scales, ranging from a simple and relatively stupid agent in a computer to a complex agent in a discipline or research program, and everything in-between. Observer systems can include measurement apparatus, sampling systems, and human collectors.

A natural classification is therefore an abstract representation that is constructed by a classifier based upon the output of an observer system in a domain in which some data is salient (Figure 1.5).

This is especially important in discussing classification, as there is a constant tendency to conflate the classification and its nomenclature, with the taxa being classified. We will later discuss this under the rubric

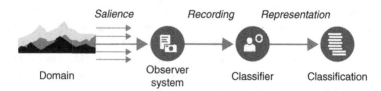

Figure 1.5 The relations between a universe of discourse or domain, and an observer system, a classifier and a classification

of the ontological fallacy in Chapter 3. The domain, observer system and classifier are physical objects, but the classification is an abstract object and so has no causal power.[30] Changes to the classification in no way affect the objects being classified. And yet, one of us observed a well meaning botanist deeply upset at the suggestion of a Phylocode[31] proponent (Brent Mishler) that we abandon the Linnaean binomials in favor of the Phylocode uninomials. "What are you doing to my plants?" he asked plaintively. Of course, nothing was being done to the plants, or indeed to the ways of studying them, but only to the naming conventions and ways of arranging the taxa. This slide from nomenclature to objects named is a well-known fallacy in logic and philosophy – the fallacy of reification or hypostasization or the fallacy of misplaced concreteness. It is ubiquitous in matters of classification, the more so since philosophy took the linguistic turn.

Notice that at no point in this sequence is anything constructed out of whole cloth. It is common to think that classification involves social construction, and of course it does, but that it involves nothing more than this, we can call pure conventionalism. In natural classification, the data are real data, the objects are real objects, and the salience may not be at all arbitrary for observer systems of the kind that we are and employ. We must look to find social aspects to classification, of course, and nothing is more relevant to this than linguistic and prior disciplinary practices, but in the end the role and justification for natural classification is that it summarizes and arranges data about the natural world rather than about the observers themselves. In sum, a natural classification ought to be a representation of the structure in real empirical data.

Notes

1. Charles Darwin to Henry Fawcett, September 18, 1861; letter 3257 of the Darwin Correspondence Project, http://www.darwinproject.ac.uk/entry-3257 accessed November 6, 2010.
2. Playfair 1802, 524f, as cited by Dott 1998, 15.
3. Best summarized in these two books: Chalmers 1990, Godfrey-Smith 2003b.
4. It must be stressed that we are discussing only scientific theories here. A wider sense of "theory" makes theory-dependence true almost by definition, but that is for discussion under more general philosophical topics of cognition and language. On models see Toon 2012.
5. Mill 1974.
6. Kuhn 1970, 205ff.
7. See the apologia for Whiggism in Mayr 1982, chapter 1, although Mayr claims not to be a Whig.
8. Erickson 2012.

9. The notion of a dynamic cognitive system is taken from van Gelder 1998a, 1998b.
10. Endler 1986.
11. Consider the periodic table: Scerri 2007. We will discuss this below.
12. The *Umwelt* of a sensory species is the aspects of its environment to which it responds and the *Umwelt* of humans is, in effect, the primate *Umwelt* of ordinary-sized objects (Griffiths and Wilkins 2013, 2013a, 2013b).
13. By "learning machine" we mean that we are what artificial intelligence researchers call "classifier systems" that take unstructured data and organize them into reliable recognition schemata (Simpson 2002). This is a result of our being neural networks, which can iteratively refine classifiers until they satisfy some function of accuracy. The best classifier system known exists between the ears of systematists and other classifiers in science (cf. Bishop 1995; Dunne 2007; Rogers and Kabrisky 1991).
14. Hacking 1983.
15. This is well discussed by Winsor 2009.
16. Described in Scerri 2007.
17. Which is to say that so long as all genders and cultures used it in the same manner, physics was doable, not that all races, cultures, and genders had equal access to the resources and opportunities needed to practice it. In other word, it was not ruled out by styles of reasoning unique to cultural and gender unprivileged classes.
18. Henrich, Heine, and Norenzayan 2010.
19. Hamblyn 2001, 103. This section is based mostly on Hamblyn's excellent book.
20. World Meteorological Organization 1975.
21. Kutzbach 1979.
22. Goethe 1825 (1970).
23. See Babich 2003 and Mößner 2011 for a discussion of their respective views and modern interpretations. Moreno 1999, and Sankey 1994, 1998 discuss incommensurability in the context of taxonomy and natural kinds.
24. Kuhn 1962. Kuhn later, as is famously known, subdivided paradigms into such entities as disciplinary matrices and exemplars under the withering analysis of Wittgenstein's student Margaret Masterson (1970; cf. Kuhn 1977, chapter 12), who noted the ambiguity and homonymity of the way he used it, in as many as 26 distinct ways. This does not affect the point here.
25. This is not to revive the logical empiricist's observation terms/theoretical terms distinction although it is not so clear that some sort of distinction must be abandoned; see Quine 1993. We are doing different things with terms that are heavily theoretically loaded than we are with terms that are more broadly observational.
26. Müller-Wille 2003, 2007; Winsor 2009.
27. Darwin 1859, 413.
28. Wittgenstein 1968, 223. If we share a form of life, as we clearly do, both being social and sexual mammals in an ecology, then we should be able to understand quite a lot of a lion's discourse. Likewise, we understand the motivations of dogs, primates and to a lesser extent cats, either because our form of life is closely evolutionarily related to them (primates) or because we have evolved in parallel fashion (dogs). *Lebensformen* are not hermeneutically

sealed from each other, contra Kuhn and Wittgenstein and many other German-Romantic-inspired thinkers. Had Wittgenstein not dismissed evolutionary biology so quickly, he may have understood this; see Cunningham 1996. Instead he reasoned from a sample species set of one.

29. Sandri 1969 gives a formalist account of this. Whether classification is a model will be discussed later.
30. Zalta 1988.
31. The Phylocode is a proposed classification scheme based upon phylogenetic trees. Ebach considers it an unstable classification system for biological systematics, as it uses phylogenetic inference (an explanandum) as an explanans (for example ancestry as evidence for a grouping). Wilkins however is sanguine, as it is in his view just another classification scheme, supposedly based on natural groupings.

2
Nature and Classification

As the science of order ("taxonomy"), Systematics is a pure science of relations, unconcerned with time, space, or cause. Unconcerned with time: systematics is non-historic and essentially static; it knows only a simple juxtaposition of different conditions of form. Unconcerned with space: geographical factors are not primary criteria in the definition of taxonomic units. Unconcerned with cause: systematics has no explanatory function as far as the origin of the system is concerned; it is merely comparing, determining, and classifying. [Thomas Borgmeier[1]]

False BOTANISTS proclaim the Laws of the Art before they have learned them:
Extol absurd Authors, and are jealous of the excellent ones:
Steal from others, producing nothing of their own:
Boast much of a little knowledge:
Pretend they have discovered a natural Method:
Assert the Genera to be arbitrary. [Linnaeus, *Supplementum Plantarum Systematis Vegetabilium* 1774, §27]

Natural system: a contradictory expression.
Nature has no system; she has – she is – life and development from an unknown centre toward an unknown periphery. Thus observation of nature is limitless, whether we make distinctions among the least particles or pursue the whole by following the trail far and wide. [Goethe, *On Morphology*, 1823[2]]

In this chapter we introduce the notion of a natural classification and the role classification plays in sciences. We consider the difference between taxonomy and systematics, and introduce the question of theory-dependence of observation. The philosophical background of classification is introduced, along with the question of essentialism and natural kinds, which we replace in classification with the method of types.

The natural system

To understand the issues in classification, it is necessary to understand some of the history and development of the debate, and from this, a kind of meta-taxonomy falls out. History explains a lot about why the issues have been framed the way they have, and recent work on the history of taxonomy explicates the philosophical issues.

At the beginning of the nineteenth century, there was a considerable debate over what counted as a "natural" classification in the sciences known then as "natural philosophy", but which had recently been dubbed "biology" by Lamarck and Gottfried Reinhold Treviranus.[3] Linnaeus' scheme for identifying and naming plants and animals had undergone a major expansion,[4] especially into the English market of plant and animal breeders,[5] and its very popularity raised a concern: how real were the groups being named? It was widely accepted that they mostly were real, despite Linnaeus himself saying that his system was artificial, that is, conventional. By the 1820s, Linnaeus' system, modified and elaborated, had become known as "The Natural System".[6] However, in France, that phrase referred to the ideas of Michel Adanson and Antoine-Laurent de Jussieu, who employed many traits as the characters by which classification was done in contrast to the Linnaean practice of employing a few "essential characters" as a definition of a species or higher group.

These two approaches to classification were held in tension as Darwin returned from his Beagle voyage, but one thing was agreed: the groups of organisms that were classified existed, and the system that Darwin called in the *Origin* "groups subordinate to groups", was a natural classification. That is, it told scientists facts about the biological world. "This classification," Darwin wrote, "is evidently not arbitrary like the grouping of the stars in constellations." It was a fact to be explained, and the explanation, which is now known as the theory of common descent, is perhaps Darwin's most original contribution to biology. Historian Polly Winsor has argued that it was natural taxonomy, not the adaptedness of organisms, that was Darwin's primary problem, which led him to the evolutionary solution.[7]

In particular, the debate over the Quinarianism of William Sharp Macleay and William Swainson involved extensive debate over what counted as a "natural" classification,[8] and these debates were in full swing as Darwin returned from the Beagle voyage to ponder his collections and their significance. While Quinarianism itself has been relegated to the dustbin of Platonist lost causes in science,[9] somewhat unfairly (it counted itself as an empirical movement), it was enormously influential on the development of the idea of natural classification in biology in general, and the discussions on the nature of classification itself were more sophisticated than credited.[10]

Elsewhere in science, classification was also being pursued actively. In chemistry there was a considerable effort to classify compounds, as we now understand them, and elements. In anthropology, of course, racial types were being classified, as well as cultures and institutions like religions. The sciences themselves were often classified, and by the end of the nineteenth century often as not classification was exemplified by a taxonomy of the sciences in philosophical texts. In all this, classification was thought to be factual.

However, by the turn of the twentieth century, classification had become a matter of convenience, of librarianship. In fact, the last extensive discussion of the matter during this period was written by a librarian.[11] Dewey, along with many others, had decided that the theory of evolution meant that it was an entirely arbitrary decision where to "carve nature at its joints", as Plato had written in the *Phaedo*, and, despite Darwin's own practice and statements, Darwin was held to have abolished the naturalness of taxonomy. In a related development, Rutherford famously declared (more than once: a good quip should be "re-used often"), "All science is either physics or stamp collecting!"[12] Classification meant arranging books, or the conceptual equivalent of books, on shelves to our own convenience. In less than a century, nature and classification had separated company. Why that is sets up the issues of this book.

On what there is

Consider this exchange:

> "Penny, I'm a physicist. I have a working knowledge of the universe and everything it contains."
>
> "Who's Radiohead?"
>
> [Long, evil pause] "Penny, I have a working knowledge of the important things in the universe." [*The Big Bang Theory*[13]]

Without stamp collecting, Rutherford knows exactly nothing about the actual universe; apart from how it would behave if there were things in it. Philosophy of science has for too long taken its lead from physics as the exemplar of scientific activity. However, physics is a general or universal science in ways that make inferences from its method and practice to other sciences, the special sciences, risky and often not well-founded. We shall return to this distinction below. In order to know something about the universe, we must first observe, and the naive and general theory-dependence of observation hypothesis falls on the problem that before you can have a theory, you must have something to theorize about. In most sciences, this is something that arises out of naive classification early in the development of the subdiscipline and the domain it investigates.

Universal languages and scientific classes

There is a traditional story told by scientists and philosophers alike, that classification begins with Aristotle, who defined kinds "essentialistically". More recently, however, this traditional story has been reexamined, and a number of things have become clear.[14] One is that Aristotle himself was not engaged in anything much like a classification of biology, or, for that matter, of scientific subjects in general. He was, rather, concerned with the logical classification of words (predicates). To some extent, he did think that we could do science by definition, at least in the logical and metaphysical works called the *Organon* (method), but this seems more like Locke's strategy of "clearing the undergrowth" than an attempt to generate knowledge by defining essences. Moreover, he did not even use a singular term for what we now call essence, instead using a phrase, the "what it is to be" some thing.[15] In his natural historical works, however, his classes of things are almost always functional classes; what we moderns might think of as models. He does not, for example, classify mammals, but instead groups some animals (whales) that are counterintuitively not animals like, say, a horse, with them on the grounds that they are red blooded, warm and give milk. This is not a taxonomy. It pays to attend to what Aristotle (and his follower Theophrastus) do differently when they do metaphysics and when they do natural history.

Aristotle, however, started a tradition, the tradition of the "ten topics". In a not-entirely-clear fashion, he suggested that all concepts could be reduced, or rather generalized, to ten ultimate concepts he called *topoi* (or "places" in the memory), which included four "predicables" (that which is predicated of things): definition (*horos*), property (*idion*), genus and accident (*symbebekos*). These predicables come in the ten classes of topoi: what-it-is, quantity, quality, relation, location, time, position, possession, doing, and undergoing.[16] While it is not clear that he intended these to be metaphysical facts, when they were mediated to the Middle Ages, Boethius, quoting Porphyry, wrote

> As for genera and species, [Porphyry] says, I shall decline for the present to say (1) whether they subsist or are posited in bare [acts of] understanding only, (2) whether, if they subsist, they are corporeal or incorporeal, and (3) whether [they are] separated from sensibles or posited in sensibles and agree with them. For that is a most noble matter, and requires a longer investigation.[17]

Once the question had been raised whether these categories and topics were real or merely in the understanding, the nominalist tradition arose not long after, relatively (around five centuries later, at the end of the so-called Dark Ages, but more importantly, when Arabic contact led to the revival of a number of classical texts and problems in Europe), and also the general view that these topics were facts about the world began to arise. Nominalists held that these general classes were simply names (hence the label, from the Latin *nomine*) given to collections of things that served our conceptions (so-called universals). This led eventually to a movement now called the Universal Language Project, which started around the time of Francis Bacon, and continued until the beginnings of modern science.[18] According to the proponents of this project, in contradiction to nominalists like Locke, there was a knowledge of the world in language, and if we could just come up with all the formally possible terms and associated ideas, we could express the nature of things clearly and without ambiguity or error. It was, in other words, the necessary precondition for doing science; first, as Cicero wrote, define your terms.[19] Locke, aware of this project, rejected the view that general terms defined "real essences", and instead famously called them merely "nominal essences". In short, universals, which we would call classes, were merely names for the convenience of communication.[20] This is also known as conceptualism, and treats classification as a merely conceptual choice.

At the same time as Locke and the Universal Language Projectors were debating, however, biological taxonomy was moving out of the morass of folk taxonomies, especially following on from the work of Andreas Cesalpino, Conrad Gesner and Caspar Bauhin in the sixteenth century. In the late seventeenth century, John Ray and his friend and colleague Francis Willughby, prepared the first florae and faunae and in the process set up the tradition of classification later codified, extended and regularized by the Swede Linnaeus and his English successors, such as John Lindley and James Edward Smith in the eighteenth and early nineteenth centuries. However, the "natural system" of the Frenchmen Michel Adanson and Antoine-Laurent de Jussieu set up a tension in the early nineteenth century. Was the Linnaean system natural or artificial? Was it forced on us by nature or a matter of convention? The Gesnerian tradition described the whole-organism, for example, while the Linnaean tradition described only those key characters in which a particular taxon differed from the others in its group.[21] This "essential definition" of the Linnaeans[22] was widely regarded as an artificial scheme when contrasted to the affinities of the Jussieu school that derived from the botany of the early Germans. Affinities were undefined, although every naturalist had an idea of what they meant by them, and included as many characters as were thought to be informative. This eventually developed into the notion of homologous relations in the mid-nineteenth century.[23]

Moreover, the fascination of scientists and philosophers for classification in the late eighteenth and early nineteenth centuries was by no means restricted to plants and animals. Linnaeus himself had attempted to include minerals under his scheme, and Mohs and others continued this tradition. Chemicals were also the subject of extensive classifications, along with diseases (nosology) from the seventeenth century onwards,[24] and so on. All these classifications were held to reflect natural states and facts.

Sometime towards the end of the nineteenth century, several developments ended this fascination for classification. One was the rejection of Baconian induction. On the older view that underpinned classificatory enthusiasm, collecting facts until they built up into a theory was thought to be a virtue of science. However, with the publication of the ideas of Herschel, Whewell and Mill, Baconian fact-collecting lost its stature, and although Whewell did argue for a "colligation" of facts as the precondition for generalizations, Mill wrote in the System of Logic that "Colligation is not always induction, but induction is always colligation".[25]

Moreover, as evolutionary ideas became widely accepted, something akin to the older *scala naturae* view that there was really nothing to tell these kinds apart became the consensus. In the first edition of the *Origin of Species*, Darwin had said "I look at the term species, as one arbitrarily given for the sake of convenience to a set of individuals closely resembling each other, and that it does not essentially differ from the term variety, which is given to less distinct and more fluctuating forms. The term variety, again, in comparison with mere individual differences, is also applied arbitrarily, and for mere convenience sake".[26] This view was widely quoted and adopted, especially around the turn of the twentieth century. In a seminar in 1908, published in the *American Naturalist*, the assumption was that Darwin thought species weren't real.[27] However, Darwin did think species were real; he just didn't think they were onto-logically qualitatively different from other groups like varieties, genera or other ranks in the Linnaean scheme. He repeatedly mentions "the subor-dination of group under group" as being explained by his theory, so he clearly thought that the arrangement was real, as well as the species.[28]

The deathblow to natural classification was provided by philosophy, and especially the views of Pierre Duhem, Henri Poincaré, and later continental philosophers, as well as the views of Bertrand Russell, the logical positivists, such as Rudolph Carnap, and Willard van Orman Quine and his followers, and the logical empiricists. These thinkers shared the view that the preeminent aspect of science was the theory, and any classification that was scientific, and not merely applied librar-ianship, would come out of theory. Modern philosophy of science tends to rely on the work of Frank Ramsey, and his invention now known as the Ramsey Sentence, a statement of the theory in first order quanti-fied logic. Since Quine's famous paper "On what there is",[29] Ramsey sentences are thought to provide the ontological classes of the domain the theory explains; as Quine wrote, to be is to be the value of a bounded variable. This was elaborated by David Lewis, in what is now called the Ramsey–Lewis Method, and more recently has been applied in the so-called Canberra Plan.[30]

This also tied into the claim by Norwood Russell Hanson that all obser-vation is dependent upon theory, and so the classes of things observed are defined by theory.[31] Philosophers as disparate as Popper and Kuhn have accepted this. It follows implicitly that classification is an operation drawn from whatever theory one has accepted, and therefore is at best a subsid-iary activity in science after the theory is developed. In his 1906 book *La théorie physique: son objet et sa structure* (*The Aim and Structure of Physical Theory* in English[32]), Duhem, a physicist, wrote that natural classifica-tion was "those ideal connections established by [the zoologist's] reason among abstract conceptions [that] correspond to real relations among the associated creatures brought together and embodied in his abstractions" [Duhem 25]. Later [Duhem 297], he says that physical theory approaches a "natural classification" by arranging experimental laws: "The ideal form of a physical theory is a natural classification of experimental laws". According to Duhem, a natural classification is a theory.

The Duhem–Quine thesis – that one cannot test a theory in isolation, but that one instead tests all theories used (confirmation holism), even if they are not directly pertinent to the phenomena being explained – would imply that classification is actually the output of all theories that could be used in a domain. No naive observation is possible.[33] Another form of this is the idea that classification is a location of an object in a space of theoretical kinds,[34] a view we might call perspectivalism. This also relates to Quine's views, although he famously allowed for evolved and innate "quality spaces" into which we might place objects. Nevertheless, this was not supposed to be a scientific way to classify.[35] If some class of objects was to be held to be real, they must be objects that occupy the conceptual buckets of our best scientific theories.

In the twentieth century philosophy of Quine, and later Lewis, classifications are simply derivations from theory. In "On what there is", Quine said that

> Our acceptance of an ontology is, I think, similar in principle to our acceptance of a scientific theory, say a system of physics: we adopt, at least insofar as we are reasonable, the simplest conceptual scheme into which the disordered fragments of raw experience can be fitted and arranged.

A classification, although Quine does not use that word, is a conceptual scheme, and, famously, "To be assumed as an entity is, purely and simply, to be reckoned as the value of a variable" [of a formal statement of a theory]. The later development of this led Lewis to use the "Ramseyfication" of theories as a way to determine what things existed under the theories, and how they related to each other.[36] Over the course of a century or so, a natural classification went from being what we observed without theory to what we derived from theory.

Things weren't helped much by Popperian insistence that theories were not inductively developed from experience and data, and that all observations were bound to theory (a view still blithely repeated by some taxonomists). So far from being the case that we developed theories by observing and classifying, as the early nineteenth century taxonomists had presumed in a naively Baconian manner, we instead could not even observe things without a prior theory. Despite Popper's view of science being largely abandoned by the mid-1970s among professional philosophers of science, it was still influential upon the new breed of taxonomists known initially as phylogenetic systematists, and later as "cladists".[37] A classification was a constructed theory, a hypothesis to be tested and falsified if possible.

What are taxa?

Homer Simpson: "Oh, I just love it here! So many things, and so many things of each thing!"[38]

The traditional approach to classification in the last century or so has assumed a kind of reductionism, in which the entities of the theory must be foundational or fundamental, even if only to the domain of discourse the theory applies over. Hence, there has been a proliferation of technical ontologies of objects such as those ending in -on, like muon, boson, electron, Jordanon, Linneon, and lastly, taxon. These are held to be either the "units of classification", or the "units of [the domain]". For example, it is often asserted that species are "the units of evolution", or of "biodiversity".[39] Attempts to remove this theory-relativity from biological taxonomy led to redefinitions of the baseline object in a "neutral" manner.[40] Each attempt was either co-opted for other purposes ("deme" was changed from a taxon to a breeding population) or it failed to replace the supposedly problematic term "species", or it faded away and the problematic term was revived. But "taxon", which is a back-formation from the term taxonomy (literally, "the rule of order"), for arranging groups in science, has persisted, and it applies to any rank or type of group that is arranged in a taxonomy. It is going to be used in this book to mean just that: something that is organized into an arrangement. A natural taxon is some object that is correctly and properly represented by a name, which denotes that object in nature, and is arranged in a natural taxonomy.

There is a long standing debate over the terms taxonomy and systematics. Systematics and taxonomy are almost, but not quite, the same thing. For example, G. G. Simpson treated taxonomy as the principles and rules of classification and systematics as the study of kinds and diversity of organisms and their relationships.[41] However, in practice either the two terms are treated as rough or exact synonyms, or they are more or less arbitrarily assigned differing meanings.[42] We need to define these terms for clarity, so throughout this book, systematics will refer to arrangements of taxa, and taxonomy will refer to the practice of describing, naming and investigating groups of organisms, following established modern practice. However, there is no fact of the matter here, and these are not, as we shall see, divided so sharply from one another.

There is considerable variation on this usage. The usual definition is something like the one proposed by Robert Ornduff:

> Taxonomy: classification of taxa (units of classification) in a system that expresses their relationships
> Systematics: comparative studies of a systematic unit (i.e., a group of organisms or species and higher), the fact-finding field of taxonomy.[43]

These seem inverted – systematics ought to be the relationships between taxa, and taxonomy ought to be the fact-finding (i.e., descriptive) side of it, but the terms were introduced that way, so that's how it got established. R. Toby Schuh[44] says that systematics and taxonomy are the same thing, and consist of three activities: recognition of species, classification into a hierarchical scheme, and placing this in a broader context. Gurcharan Singh also treats them as identical activities.[45] Richard Mayden does not discuss taxonomy, but defines systematics as "the field of science concerned with reconstructing the evolutionary or ancestor-descendant relationships of groups of organisms, whether fossil or recent, on the basis of heritable traits".[46] Overall, there is much confusion, as summarized by Peter F. Stevens:

> Words like "method," "system," and "systematics" are perhaps the key words used by [his subjects for the book], and I must clear up some of the ambiguities surrounding their use. First, as to the distinction between taxonomy and systematics, Simpson offered a much-quoted definition of systematics: "the scientific study of the kinds and diversity of organisms and any and all relationships among them." Classification was the grouping of organisms into the hierarchy of a classification; taxonomy was the theoretical study of classification. For Frans Stafleu, on the other hand, taxonomy was represented by keys, systematics by interpretive relationships. Recently, a different distinction has been drawn between systematization and classification, the former being an ordering according to element/system or part/whole relationships, the latter of categories based on common properties.[47]

The terms were defined independently of each other, by Augustin P. de Candolle for taxonomy, 1813, and John Lindley for systematics,

1830.[48] De Candolle defined classification as having three components in his *Théorie élémentaire de la botanique*: Glossologie (we would call it nomenclature now), Taxonomie (the theory of classifications), and Phytographie (the rules of describing plants).[49] Lindley, an adherent of the Jussieu scheme of multiple lines of evidence rather than single keys in the Linnaean system, used the term "systematic botany" for this approach, which he thought was a natural classification system in contrast to the artificiality of Linnaeus. He did not use the term "systematics" directly.

What is significant with de Candolle's scheme is that he includes arbitrary and natural facets under the single heading of "classification". Rules of description, or phytography, are roughly the same as what a modern systematist would mean by "taxonomy": the description of species from specimens, and what he would mean by taxonomy is what we would now mean by systematics: the arrangement of the species into schemes of relationships. A student of both de Candolle's approach and of Lindley's in botany, was the very influential Asa Gray. His influence is threefold. First, he was American, and hence influenced the later generations of American botanists and their theoretical ruminations. He is quoted as late as 1935 as an authority on just this matter in the Proceedings of the Linnean Society of London.[50] Second, he was a botanist, and the botanical discussions ran rather parallel with and in some cases in contradiction to the zoological ones. And third, he was a Darwinian, and so his strictures were regarded as modern enough to accept.

Gray adopts de Candolle's overall view, despite the fact that classification is by then thought to be explained by evolution. In his *Structural Botany*, Gray distinguishes between

> Taxonomy, or the principles of classification, as derived from the facts and ideas upon which species, genera, &c., rest; Classification or the System of Plants, the actual arrangement of known plants in systematic order according to their relationships...[51]

as well as several other aspects of "Systematic Botany" such as Phytography (rules for description), Glossology or Terminology, and Nomenclature (the methods and rules adopted for the formation of botanical names). Here, systematics includes both taxonomy and classification. This can be shown as a diagram (Figure 2.1).

Gray thought, as most systematists did at the time, that the taxonomic and classificatory aspects were in fact dealing with natural truths,

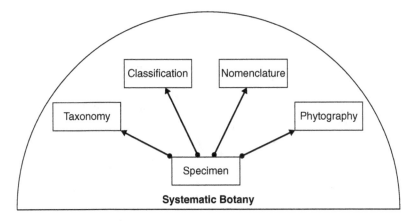

Figure 2.1 Asa Gray's scheme of classification

the former from a theoretical or methodological point of view, the latter from an empirical or epistemological point of view. The communicative and conventional aspects of nomenclature and phytography, respectively, were not natural. We might generalize this for all scientific classificatory activities thus (Figure 2.2).

This is eminently sensible, so why is it not currently the default view? A lot happened in the twentieth century, and not all of it was to do with Julian Huxley's New Systematics.[52] In a symposium in 1935, held by the Linnean Society of London, Walter B. Turrill introduced the terms "alpha" and "omega" taxonomy, and it caught on. As he defined it,[53]

> ...the time has come when the student of floras whose taxonomy on the old lines is relatively well known should attempt to investigate species by much more complete analyses of a wider range of characters than is now the rule. There is thus distinguished an alpha taxonomy and an omega taxonomy, the latter being an ideal which will probably never be completely realized. ... The aim of the alpha taxonomist must be to complete the preliminary and mainly morphological survey of plant-life ... Some of the criteria which those who aim at an omega taxonomy are ... ecological, genetical, cytological, and biometrical.

So, alpha taxonomy became exclusively associated with species-level work, thereby conflating the older term for all of systematic biology. But the way Turrill introduced this term indicates that he saw the omega

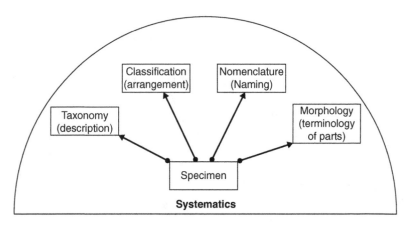

Figure 2.2   A general scheme of classification

taxonomy as the final form of all classification, along some continuum of completeness and naturalness.

As the field changed under the influence of Simpson and Mayr's insistence that taxonomy/systematics were the same, and that they were aimed solely at the reconstruction of evolutionary history (the view that Mayden advocates), two further developments occurred. One was the numerical taxonomy movement started by microbiologist Peter H. A. Sneath and mathematician Robert R. Sokal.[54] Here several issues were in play. They wished to make classification non-subjective (that is, objective) and take the decisions out of the hands of biased observers. To do this they introduced mathematical algorithms that could be implemented in computer programs or done by hand. They relied upon any data whatsoever, without prior filtering, in order to achieve naturalness. Another issue was that they wished to make the process of classification purely operational, following Percy Bridgman's philosophy that all that counted in science was how things were measured (operations of measurement, hence the name operationalism[55]). The problem here that arose was that depending upon the principal components chosen, different taxa fell out. While numerical taxonomy, which came to be known as "phenetics" (from the Greek *phaneros*, apparent, manifest) found structure in the data, it seemed that structure was not always, nor even often, taxonomic structure.

Still, Sneath and Sokal's books were extremely clear and coherent, and they set up the contrasts in modern taxonomy, which was that taxonomy and systematics were the same thing. Around this time also, Hennig's *Phylogenetic Systematics* was published in English, along with Lars Brundin's classification of midges using Hennig's method,[56] leading to the school[s] of thought now called (following Mayr's reading of Rensch, and Cain and Harrision) cladistics.[57] On this view, initially, classification was solely aimed at reconstructing evolutionary history (which was something Hennig appealed to Simpson and Mayr to support). Later critics argued that classification did not give the history but that the history was an inference or hypothesis based upon and tested by the classification, which was independent.[58] The history view was confirmed for philosophers by Elliot Sober's very influential *Reconstructing the Past.*[59]

Finally, another school arose, one which as yet has no simple name. On this view, classification is dispensed with entirely, and systematics is only about phylogenetic reconstruction, usually employing statistical methods of analysis of very large and mostly molecular data sets. The champion of this view, both theoretically and practically, is Joe Felsenstein, whose book *Inferring Phylogenies*[60] is now the standard bible for methods and algorithms, as well as espousing what he calls the "It Doesn't Matter Very Much" school of classification. Felsenstein also maintains a site from which you can download nearly every computer program used in systematics. He holds that classification, and the philosophy that underpins it, is a matter of personal preference, and nothing much hangs on what one chooses:[61]

> A phylogenetic systematist and an evolutionary systematist may make very different classifications, while inferring much the same phylogeny. If it is the phylogeny that gets used by other biologists, their differences about how to classify may not be important. I have consequently announced that I have founded the fourth great school of classification, the It-Doesn't-Matter-Very-Much school.

With this, the assimilation of systematics into phylogenetics appears complete. All is now subordinated to the finding out of historical pathways that explain our present biodiversity. We shall call this *statistical cladistics*. What counts is the use of phylogenetic analytical techniques. However, phylogenies are, if not classifications, nothing much at all (we shall consider the claim they are histories in Chapter 7). In fact, classifications take priority over phylogenies, but that is a much more radical argument for later in the book. If statistical cladistics is not classification, it is hard to see what it might otherwise be. It is, in effect, classification without classes.

What are the relations of taxa to their individual items?

The traditional essentialist story, which is itself a construct of the mid-twentieth century, was devised to set Darwin out from his contemporaries for the centenary of the *Origin* in 1959. According to this story, a taxon is supposed to be a class before Darwin, and its members are all the objects that have a shared essence, or set of unique properties. After Darwin we are now able to see that taxa are populations, and that the relevant relation of the member to the taxon is one of inclusion rather that instantiation. That is, a member of a species, for example, is a part of the historical object that is the species in time and space. Previously, a member of a species was just an instance of the class "species *A b*".

That this is historical invention is irrelevant to the issues, for it is representative of the contrasts employed by those working in the post-1959 era, and recently, Michael Devitt among others has attempted to "revive" (that is, establish) an essentialist account of taxon membership.[62] This is not quite the supposed Platonism of the pre-Darwinians: Devitt does not suppose that the form of the taxon exists in some Platonic realm, while the members are only approximations of the Truth, a view only ever held by Louis Agassiz amongst naturalists so far as we can tell.[63] Instead Devitt suggests that every member of a species must have some set of causal and intrinsic properties to be a member of the species. This is more Aristotle than Plato: the taxon is not quite real, while the organisms and their properties are. Concerning essences in (logical) taxa, John Locke wrote in the *Essay*

> The learning and disputes of the schools having been much busied about genus and species, the word essence has almost lost its primary signification: and, instead of the real constitution of things, has been almost wholly applied to the artificial constitution of genus and species. It is true, there is ordinarily supposed a real constitution of the sorts of things; and it is past doubt there must be some real constitution, on which any collection of simple ideas co-existing must depend. But, it being evident that things are ranked under names into sorts or species, only as they agree to certain abstract ideas, to which we have annexed those names, the essence of each genus, or sort, comes to be nothing but that abstract idea which the general, or sortal (if I may have leave so to call it from sort, as I do general from genus), name stands for. And this we shall find to be that which the word essence imports in its most familiar use.[64]

Locke is discussing the way in which the scholastics, as we now call them, treated general taxonomic terms (*genera* and *species* in Latin) as abstract ideas, or as names of abstract ideas. Given Locke's nominalism, it is unsurprising that he thought these were "nominal essences" only, but the nominalist position is not, of course, the only one. The Platonist view, held only by Agassiz, is that organisms (and other objects) of a taxon are neither parts of the taxon nor instances of the taxon (tokens of the type), but are at best approximations of it. The taxon is quite literally a platonic form, and objects are not exemplars of that form except in a rough way. Moreover, if there were no tokens of the form, the form would remain as an ideal object; in philosophical Latin, the form exists *ante rem*. Platonism is rare in natural science, but the temptation to treat taxa as abstract objects of this kind is constant.[65] The other traditional position here is that the physical objects instantiate the form, and the form would not exist without at least one of these objects, but the form is distinct from the physical aspects of the objects. This is, of course, Aristotelianism, and the Latin is that the form exists *in rebus* (in things).

Aristotelian taxa are based on a form/matter distinction known as hylomorphism, from the root Greek words for "matter" (*hûlos*) and shape or form (*morphê*). It was set up against atomism and Platonism, and relies on there being some causal role for form in scientific explanations. Atomists, now often referred to as reductionists or physicalists, hold that the properties of any object arise from its constituent physical parts, bottom-up. As a result, physicalists tend to be nominalists about taxa and "universals".[66] For hylomorphists, though, the relation between the physical object and its taxon is one of top-down causation. The form, the taxon, causes the object to be what-it-is. This is a very common idea in biology, but less so in other sciences. However, recently a form of hylomorphism has developed based on information instead of form. Informational accounts of genes, development, cognition, and even physics itself, have become, if not popular, more widely made out.[67] According to this view the taxon is partly composed of informational properties as well as physical ones.

Hylomorphism is based on an analogy between the potter and the clay, and is itself a methodological creationism: things would not otherwise naturally be what they are without intelligence or design imposed upon them.[68] It strikes us as odd that it would persist today, when practically every claim made by the atomists has been borne out. In any case, in the present context, form is crucial in classifying not because it is something that causes things to be in their taxa, but because it diagnoses them. In fact, it constitutes being taxa and taxonomies. This leads to the question of what kinds of kinds exist.

Natural kinds

6.3431 The laws of physics, with all their logical apparatus, still speak, however indirectly, about the objects of the world. [Wittgenstein, *Tractatus*[69]]

The traditional way of handling natural classes of things in the philosophy of science has been to treat them as natural kinds.[70] Natural kinds are classes with shared properties but more than this philosophers of science do not agree upon. The situation is nicely summarized by Ingo Brigandt:

The traditional account (used especially by metaphysicians and philosophers of language) construed a natural kind as a special type of class characterized by two features. (1) All members of a natural kind have the same characteristic properties, permitting universal generalizations, such as laws of nature (e.g., all oxygen atoms share physical properties and can undergo the same chemical reactions). (2) The identity and boundary of a natural kind is metaphysically determined by an essence; an object belongs to the kind in virtue of having this essential property. The essence is epistemologically fundamental in that it explains the characteristic properties of the kind (e.g., the essence of oxygen is its atomic structure, which explains all physical and chemical properties of oxygen).[71]

As Brigandt goes on to note, this traditional, Millian, notion of a natural kind has been supplemented or replaced for biological taxa by a notion of Richard Boyd's,[72] the homoeostatic property cluster (HPC) view of kinds. According to this approach, taken by a number of philosophers with respect to species,[73] what makes a natural (biological) kind is that a number of properties cluster in such a way as to maintain the kind causally. While this may work at the level of a contiguous population, and possibly therefore a species, it does not apply to higher taxa,[74] necessitating a different approach to higher taxonomic essences. Paul Griffiths proposed that the essence of a taxon is historical, a shared ancestry,[75] but while this may in fact be the case, it only applies to monophyletic taxa,[76] and excludes taxa formed by shared properties irrespective of history; and in any case, the reason history is important is because of retained properties (synapomorphies, a term we shall later discuss in detail). The historical essence view also has epistemological problems – we do not necessarily know that two taxa share a common ancestor, and so while it may be an explanation of the synapomorphies that constitute a taxon, shared ancestry cannot be the reason why we group the taxon thus, since it is inferred, rather than observed.

There are a number of kinds of essential kinds.[77] The most obvious we shall call definitional (Ciceronian or conceptual) kinds. Cicero was fond of "defining terms" before investigation began, and this has become something of a habit amongst philosophers. However, while everybody knows that mere definition is not science (although it is an insult to linguists to dismiss this as "mere semantics"; semantics matter tremendously), definition of ordinary language terms, for example by Aristotle, who we might think of as the first Ordinary Language philosopher,[78] is often thought of as a way to gain knowledge about the world. This can be criticized as an inversion of true empiricism, and often it is, but there may be some justification for it in the degree of vernacular knowledge embedded in a language community. Still, Aristotle's attempt to do science by definition is matched also by his attention to empirical detail, and in our view, he only does the former when doing what we might now think of as conceptual analysis rather than natural history.[79]

Another kind of kind is the productive kind. This was basically how kinds were conceived of before the rise of heredity as a field of scientific study in the nineteenth century. A kind of animal or plant was a kind that produced more of itself, through generative powers that cause the progeny to resemble their parents. One of us has already argued that this is a generative conception of species in biology.[80] In sciences more generally, a productive kind is any kind that causes more instances of itself to occur.

The third kind, however, is not truly essentialistic. There is a history of philosophers using biological examples when discussing kind terms, usually in the context of definitions in logic, but until the nineteenth century, few philosophers closely attended to the practices and notions of taxonomy. William Whewell is particularly interesting, as he effectively asserts that what is going on in classification is rather more like Wittgenstein's family resemblance predicate than the modern account of natural kinds. He writes, in volume 2 of the *Philosophy* which deals with the life sciences, opposing the definitional tradition of species in logic:

> But it may be asked, if we cannot define a word, or a class of things which a word denotes, how can we distinguish what it does mean

from what it does not mean? How can we say that it signifies one thing rather than another, except we declare what is its signification? The answer to this question involves the general principle of a natural method of classification which has already been stated and need not here be again dwelt on. It has been shown that names of kinds of things (genera) associate them according to total resemblances, not partial characters. The principle which connects a group of objects in natural history is not a definition but a type. Thus we take as the type of the Rose family, it may be the common wild rose; all species which resemble this flower more than they resemble any other group of species are also roses, and form one genus. All genera which resemble Roses more than they resemble any other group of genera are of the same family. And thus the Rose family is collected about some one species which is the type or central point of the group.

In such an arrangement, it may readily be conceived that though the nucleus of each group may cohere firmly together, the outskirts of contiguous groups may approach, and may even be intermingled, so that some species may doubtfully adhere to one group or another. Yet this uncertainty does not at all affect the truths which we find ourselves enabled to assert with regard to the general mass of each group. And thus we are taught that there may be very important differences between two groups of objects, although we are unable to tell where the one group ends and where the other begins; and that there may be propositions of indisputable truth, in which it is impossible to give unexceptionable definitions of the terms employed.

15. These lessons are of the highest value with regard to all employments [*sic*] of the human mind; for the mode in which words in common use acquire their meaning, approaches far more nearly to the Method of Type than to the method of definition.[81]

Whewell is arguing initially that unscientific terms lack definitions, but he extends this a page later to natural history:

We may further observe, that in order that Natural History may produce such an effect [to correct "the belief that definitions are essential to substantial truth"], it must be studied by inspection of the objects themselves, and not by the reading of books only. Its lesson is, that we must in all cases of doubt or obscurity refer, not to words or definitions, but to things. The Book of Nature is its dictionary: it is there that the natural historian looks, to find the meaning of the words which he uses.... So long as a plant, in its most essential parts, is more like a rose than anything else, it is a rose. He knows no other definition. (519–520)

Whewell is merely repeating the views widely held by naturalists since Linnaeus.[82] As someone once noted, if Whewell describes it, then it is the received view of the day. Systematists had long held to something like a plenum view, as Stevens has shown, in which there were an increasing number of empty "spaces" as time went on and more species were described, leaving unfilled intermediates. By the early 1830s, the notion that there was a type species in every genus, and a type genus in every family, was common. Mill himself says as much in the *System of Logic*.[83] So by the time Mill writes, the method of type is well established. Mill is the one making a novel account, not Whewell. Mill has, in effect, changed the subject.[84] Whewell's kind is an exemplary kind, not unlike the notion of a prototype in modern philosophy of language.

The publication and acceptance of evolutionary theory revived a view of classification as arbitrary divisions of the plenum that had preceded Linnaean taxonomy. Linnaeus himself had written that genera and species were the work of nature, and all else was the work of the mind, and Buffon, among others, had held that in nature there were only individual organisms, and all taxonomic categories were merely conveniences. Once Darwin's ideas on gradual evolution of species, and on phylogeny, as Haeckel termed the idea of branching descent, were accepted, many immediately retreated to the view that this meant that any classification, even at the species-level, was mere convention. Wilkins has argued against this interpretation of Darwin in his history,[85] but few after Darwin doubted that the species concept, and all classification, was arbitrary. Haeckel, and the German systematists who followed after him, were the exceptions.

The problem with an evolutionary, or in more general terms, a process account of kinds, is that if the historical process can move in small steps from one taxon to another, it appears that taxa are matters only of convenience to the specialists. This applies in any historical science, including Earth sciences, sociological sciences, and even such fields as astronomy and archeology. Whewell called these the palaetiological sciences

The sciences which treat of causes have sometimes been termed ætiological, from αἰτία, a cause, but this term would not sufficiently describe the speculations of which we now speak; since it might include sciences which treat of permanent causality, like mechanics, as well as inquiries concerning progressive causation. The investigations which we now wish to group together, deal, not only with the possible, but with the actual past; a portion of that science on which we are about to enter, geology, has properly been termed palæontology, since it treats of beings which formerly existed. Hence combining these two notions, the term palætiology appears to be not inappropriate, to describe those speculations which thus refer to actual past events, but attempt to explain them by laws of causation.[86]

The palaetiological sciences (to modernize the spelling) lead us to the problem of temporal kinds. There have been two ways to deal with them. One we call gradism, from the ideas employed by the evolutionary systematists; the other we call cladism. Gradism is the view that what matters in a classification is the degree of difference that evolution or other palaetiological processes produce. Cladism is the view that degrees of difference are subjective, and that what counts are differences of identity, which we shall consider later.

Many after Darwin held that because evolution was a gradual process, there was no fact of the matter about when species changed, and because such things were Sorites heaps (that is, like the properties of being bald or hairy, they had no sharp boundary between them), the things themselves were not real, and were merely a matter of convenience in communication. That was the equivalent of saying that they were not natural objects. So, the problem arose of what it was in evolution that did evolve and was natural. Attempts were made to stress the discontinuity between species, both at a single moment and over time (that is, synchronically and diachronically). Mayr, for example, would assert that species were "non-dimensional", by which he meant they existed solely at a single moment in relation to other species. George Simpson, on the other hand, held that species were processes that occurred over time and were evolutionarily distinct from each other in their fate. This led him to propose that what evolved were "lineages" of populations.

Lineage is a basic concept required for temporal kinds, and in biology means a sequence of populations that give rise to each other, or a sequence of parent-child relations between organisms (see the notion of a "tokogenetic lineage" in Hennig's work), or even a sequence of genomes. "Lineage" is a substrate-neutral concept, that can be applied whenever there are the requisite causal reproductive relations between predecessors and successors. For example, Kevin de Queiroz has proposed a "Metapopulation Lineage Concept", in which a species is a sequence of lineages of a single metapopulation. The notion of a lineage plays a crucial role in the work of Hull on the metaphysics of species.[87] Lineage is a very rough concept. In the sciences generally, a lineage is best thought of as a causal chain of reproducers,[88] a sequence of causes and effects, or, as the term is employed in the biomedical sciences, an etiology. Etiological classifications are the basis for scientific explanations based on taxonomies.

Types and taxa

A distinction is sometimes made between typology and taxonomy, for instance in the title of a book on social science methods.[89] Different disciplines, ranging from archeology to computer science, apply the terms typology and taxonomy in different ways. In some, they are roughly the same. In others, a typology is a mere intellectual construct and a taxonomy an empirical arrangement. Sometimes typology is a subset of taxonomy, especially in the recent appropriation (and general inversion[90]) of these terms by computer systems analysts. The standard view held by biological taxonomists is that taxa are singular objects while types are classes, and typology is regarded as a regressive and pre-evolutionary approach to the data and biology. An early criticism of cladistics by Mayr was that it was, he dismissively wrote, "typological", and hence not properly evolutionary. In reply, Hennig, the "father" of phylogenetic systematics, accused Mayr himself of being "Aristotelian" and "typological".[91] Such name calling is unhelpful in understanding the issues. For a start, there is a clear difference between one of Mayr's named taxonomic sins – essentialism – and the other – typology – that he failed to grasp, instead treating them as synonyms.[92] More recently, Winsor has noted that systematists both before and after Darwin used a "method of exemplars", or as Whewell called it, a method of types, to arrange their taxa.[93] Types are crucial in most natural classification because they are the phenomena around which classifications are made. The basal taxa, the least inclusive groupings, of classification are usually types in many sciences, especially the historical sciences like geology, astronomy, biology and the social sciences. By conflating "type" and "essence", Popper and Mayr have led many to conclude that types are artificial and arbitrary, and only individuals, in the metaphysical sense, exist. We shall return to this point in more detail later.

Types are the foundation of taxonomy. A taxonomy arranges taxa, which are identified as types formally prior to the taxonomizing, although of course neither the types nor the taxonomies are commenced upon tabula rasa. Without types, no taxa could be compared, let alone arranged, as no taxa could be identified. However, it is an open question whether types exist as conceptual constructs alone, or as objects in the world independently of how we conceive of them. For a natural classification to be made, the phenomena that the types summarize or collect around must be real,[94] even if once the phenomena are explained in theoretical terms, they are divided differently to the initial typology.

In using the term type, some distinctions are required. In table 2.1 we list known definitions of various kinds of type terms in mostly biological and philosophical literature. We propose to define mesotype as the exemplary or emblematic instance of a type (from "middle form"). Mesotypes are crucial in the identification of types, and of subsequent classification. However, as has been widely discussed since the inception of genetics in biological systematics, a mesotype is not an essence, since most populations have a distribution of types (genotypes) and often have multimodal distributions. Taxonomy in general relies deeply upon identifying types in order to group and arrange organisms; elsewhere one of us has called this the "method of type" following Whewell.[95] We therefore reject the dismissal sometimes made of this or that philosophical approach to classification as being "mere typology" or "typological" as if that were sufficient to relegate it to the errors and shortcomings of the past. The question is instead, what types do we employ in classifications, and are they natural objects?

Some of this antipathy to types has arisen from the political infighting of systematists, a field which has been marked by its strong and personal disputes over the past few decades. Mayr, as noted, conflated types with essences, and attached both ideas to a constructed Aristotle. Those who wished to emphasize the modernity of their own view would insist upon it being in line with either Darwin, or if Darwin was not modern enough, current evolutionary biology, and relegate "typology" to an outmoded metaphysics. Given that this metaphysics never actually operated in biology, this looks more now like scientific political special pleading than argument. It is time to reevaluate.

Types, like any other general category, come in types. A taxonomic type is a distinct thing from an anatomical type, a genetic type (genotype), or a developmental type. This book is predicated on the assumption that taxa are types of things that get classified in a domain, however, and so what is a part in one domain of classification (biological systematics, for example) may be a taxon on another domain (anatomy). This level-relativity is crucial in understanding how types apply. The usual presumption for a practitioner in one domain is that all other types are derivative of that privileged domain. This need not be so, and therefore we should implicitly or explicitly track the domain by indices: this is a type[biol] and that is a type[anat], and so on. Sometimes a type of organ or other anatomical structure supports a type of taxon; sometimes it does not. When it does, is the topic of the chapter on homology and analogy (Chapter 4).

Infimae taxa and ranks

In the traditional logic that developed out of the medieval logical tradition, an infimae species was the smallest logical or semantic class, divided out of a higher class called a genus, which itself was a species of a higher genus until one reached the highest genus, called exactly that in Latin: summum genus. An infimae species was the smallest group and contained only individuals, which meant a metaphysical individual, something that had a proper name like Plato, or Aristotle.

In natural classification, it is often thought that there has to be an infimae taxon. The term taxon has no set rank or scale, and can apply, like genus, to groups of taxa. A classification, however, arranges taxa, and so there needs to be something in each classification that is duly arranged. In biology, this gives rise to the species problem: what are the units of evolution, or classification (or ecology, geography, and so on)? Much ink has been spilled on this problem.[96] If species, and by extension any infimae taxon for any natural discipline, are the units of a theoretical account, then they are explanations,[97] but if they are, as we argue here, the outcome of observational salience, they are phenomena that stand in need of explanations; they are explananda.

However, it is not necessary that an infimae taxon be defined or recognized before classification can be done. We do not need "units" to classify – specimens are sufficient. A classification can be done, and ideally is done, on the results of observation of objects. Taxa are then constructed as we begin to recognize patterns in the data. A cladogram, for example, may be constructed when all the classifier has are individual specimen fossils that have no other instances. The tokens stand as type specimens, however, so how are taxa recognized as patterns in these cases of a singular token? The solution is that patterns are relative relations, and so we recognize the specimen as a "different type" only because we already have prior knowledge of things that are in relationship to it and identify that it does not fit neatly into the patterns they generate. Hence, there is a pattern, but it is a pattern of exclusion: the new taxon is formed from the joint assumption that the specimen must reside in a taxon, and that it does not reside in existing related taxa. Specimens stand as surrogates for types. The observer system does not see the types directly, but the tokens generate data patterns that the observer system is inclined to group or divide. We do not go into the field and see species, but neither do we go into the field armed with a theoretical ontology that permits us to see specimens as tokens of species. Rather, the act of seeing forms phenomena from specimens and relationships we already know. There is no shallow classifying perceptual module – just an informed and trained taxonomist. The infimae taxa are formed by the act of observing and the experience of past observations. However, having said that, we should not infer that theory never modulates observation; of course it does. There is always a feedback on observation from prior knowledge and experience. However, if no prior experience or knowledge speaks to the domain under investigation, it cannot modulate that investigation.

Table 2.1   Type terms in the *International Code for Zoological Nomenclature* (ICZN), the *International Code for Botanical Nomenclature* (ICBN), *The Oxford English Dictionary* (OED), history of biology, and philosophy

Types	Definition
Type	"The general form, structure, or character distinguishing a particular kind, group, or class of beings or objects; hence transf. a pattern or model after which something is made ... A kind, class, or order as distinguished by a particular character ... A person or thing that exemplifies the ideal qualities or characteristics of a kind or order; a perfect example or specimen of something; a model, pattern, exemplar. ... Nat. Hist., etc. A certain general plan of structure characterizing a group of animals, plants, etc.; hence transf. a group or division of animals, etc., having a common form or structure ... Nat. Hist. A species or genus which most perfectly exhibits the essential characters of its family or group, and from which the family or group is (usually) named; an individual embodying all the distinctive characteristics of a species, etc., esp. the specimen on which the first published description of a species is based." OED
Allotype	"A designated specimen of opposite sex to the holotype. This term has no name-bearing function and is not regulated by the code (Rec. 72A)". ICZN

Continued

Table 2.1 Continued

Types	Definition
Antetype	"A preceding type; an earlier example". OED
Antitype	"One of the opposite or contrary kind". OED
Archetype	The original from which all others are copied (sensu Owen).
Autotype	"A reproduction in facsimile". OED
Biotype	"A group of organisms having a common genotype". OED
Collotype	"A thin plate or sheet of gelatine, the sensitized surface of which has been etched by the action of the actinic rays, so that it can be printed from; also the print or impression, and the process". OED
Cotype	"an additional type-specimen". OED
Ecotype	"A phenotype of a species which is the result of genetic or developmental adaptation to life in a specific environment". OED
Ectype	A reproduction or copy, as contrasted with an archetypal original.
Epitype	"A specimen or illustration selected to serve as an interpretative type when the holotype, lectotype, or previously designated neotype, or all original material associated with a validly published name cannot be identified for the purpose of precise application of the name of a taxon (Article 9.7)." ICBN
Ex-type	"A living isolate obtained from the type of a name when this is a culture permanently preserved in a metabolically inactive state (Rec. 8B.2)." ICBN
Genotype	The genetic composition of an organism or of groups of organisms.
Haplotype	"A species taken as the type species of a genus because no other was originally included in the genus". OED
Hapanotype	"A special kind of holotype in the case of extant protistans, which can consist of more than one individual (Article 73.3)". ICZN
Holotype	"A single specimen designated or otherwise fixed as the name-bearing type of a species name when it was first described (Article 73)". ICZN
Homotype	Serial homologues; repeated or serially related parts. Owen 1848, 8, Williams, Ebach, and Nelson 2008, 134
Isotype	"A type or form of animal or plant common to different countries or regions" or " Bot. A duplicate of the holotype)" or "Any mineral which is isotypic with another; an assemblage of minerals of which all the members are isotypic with one another". OED
Isosyntype	"A duplicate of a syntype". ICBN
Karyotype	"The chromosomal constitution of a cell (and hence of an individual, species, etc.) as determined by the number, size, shape, etc., of the chromosomes (usually, as observed at metaphase during cell division)." OED

Continued

Table 2.1   Continued

Types	Definition
Lectotype	"One of a number of syntypes which has been designated later as the single name-bearing type of a species, the remaining syntypes become paralectotypes and have no further name-bearing function (Article 74)" ICZN; "A specimen or illustration designated from the original material as the nomenclatural type if no holotype was indicated at the time of publication, or if it is missing, or if it is found to belong to more than one taxon (Article 9.2)." ICBN
Monotype	"A genus or other taxon based on a single species or specimen" or "The sole species of a genus or other taxon" or "Virol. A strain within a single serotype of a rotavirus, distinguished by monoclonal antibody neutralization". OED
Neotype	"A single specimen designated as the name-bearing type of a species name when the original type(s) is lost or destroyed and a new type is needed to define the species. Under exceptional circumstances the Commission may use its plenary powers to designate neotypes for example if an existing name-bearing type is not in accord with prevailing usage (Article 75)" ICZN; "A specimen or illustration selected to serve as nomenclatural type if no original material is extant or as long as it is missing (Article 9.6)." ICBN
Nomenclatural type	"The element to which the name of a taxon is permanently attached (Article 7.2)." ICBN
Paratype	"Where there is a holotype, the other specimens in the type series are paratypes (Rec. 73D), and they have no name-bearing function". ICZN; "A specimen cited in the protologue that is neither the holotype nor an isotype, nor one of the syntypes if two or more specimens were simultaneously designated as types (Article 9.5)." ICBN
Paralectotype	See lectotype; ICZN
Phenotype	"The sum total of the observable characteristics of an individual, regarded as the consequence of the interaction of the individual's genotype with the environment; a variety of an organism distinguished by observable characteristics rather than underlying genetic features". OED
Polytype	"A polytypic form of a substance". OED
Pretype	"To prefigure; to foreshadow". OED
Prototype	"The first or primary type of a person or thing; an original on which something is modelled or from which it is derived; an exemplar, an archetype." OED
Proxytype	"can be a detailed multimodal representation, a single visual model, or even a mental representation of a word". Prinz 2002, 149

Continued

Table 2.1 Continued

Types	Definition
Serotype	"A serologically distinguishable strain of a micro-organism". OED
Somatotype	"The physique of an individual as expressed numerically in terms of the extent to which it exhibits the characteristics of each of three extremes (the endomorph, mesomorph, and ectomorph)." OED
Stereotype	In Putnam's philosophy, the typical surface phenomenon that terms attach to. A "normal member of a natural kind". Putnam 1975, 148
Subtype	"A subordinate type; a type included in a more general type; spec. a subdivision of a type of micro-organism". OED
Syntype	"Where a description has been based on a series of specimens, these collectively constitute the name-bearing type (Article 72.2–72.4; 73.2)." ICZN; "Any specimen cited in the protologue when there is no holotype, or any of two or more specimens simultaneously designated as types (Art. 9.4)." ICBN
Topotype	"A specimen from the locality where the original type-specimen was obtained". OED

ICZN online http://iczn.org/content/what-kinds-types-are-there online, accessed December 29, 2012.
ICBN http://ibot.sav.sk/icbn/main.htm online accessed December 29, 2012.
Oxford English Dictionary online, accessed July 20, 2010.

Notes

1. Borgmeier 1957.
2. Goethe 1988. Goethe's contribution to the science of observation is highly significant, but is not covered in this volume. Instead we refer the reader to Brady 1972, 1983, Bortoft 1996, Seamon and Zajonc 1998, Steigerwald 2002 and Ebach 2005.
3. However, an earlier use by Hanov predates either by 30 years (McLaughlin 2002).
4. Cain 1995; Cain 1959; Müller-Wille 2003, 2007.
5. Pavord 2005, 2009 gives the pre-Linnaean history of taxonomy accessibly. Yoon 2010 equally accessibly covers the post-Linanean history.
6. "Natural" had a different signification in France, where it meant "agreeable to the mind".
7. Winsor 2009.
8. Macleay 1819; Swainson 1834, 1835.
9. Coggon 2002; Hull 1988, 92–96; Mayr 1982, 203; Panchen 1992, 118–121.
10. Although Quinarians did behave like the Hullian systematists-at-war, perhaps leading to their unpopularity.
11. Richardson 1901.

12. Johnson 2007 is an excellent survey of the history of this phrase and the attitudes to "stamp collecting" in the sciences.
13. "The Work Song Nanocluster", 2009.
14. Anderson 2005; Atran 1990; Charles 2002; Stevens 1994; Wilkins 2010; Winsor 2003, 2006a. Wilkins summarizes the debate and issues in his Wilkins 2013, 2013a, 20132b.
15. τό τί ἦν εἶναι, e.g., Top. A.5, 101b39, E.3, 153a15–16, Met Δ.6, 1016a33 (Baum 2009). The term "essentia" was a Latin back-formation; "to-be-ness".
16. Topics 103b22. See page 8 of Aristotle and Smith 1997.
17. Anicii Manlii Severini Boethii In Isagogen Porphyrii commenta, editio 2a, lib. I, ca. 10–11, Samuel Brandt, ed., p. 159 line 3 – p. 167 line 20. Translation by Paul Spade.
18. Wilkins covers the history of this in his 2009b. See also Rossi 2000; Slaughter 1982; Wilkins 1668.
19. Tusculan Disputations, Bk V, §25.
20. *Essay on Human Understanding*, Bk II, chapter 32, §6; Bk III, chapter III, §15; Bk III, chapter V, §9; Bk III, chapter 6, §25.
21. With the help of Thony Christie, Wilkins was able to examine Conrad Gesner's unpublished drawings (1555–1565) for his *Historia Plantarum*, which he died before completing, at the Universitätsbibliotek Erlangen-Nürnberg. These beautiful diagrams showed multiple stages of the lifecycle of the plant, and used whole-organism characters for identification, unlike the Linnaean arbitrary use of sexual organs only, and hence at the mature stage of the developmental cycle.
22. This is the *character essentialis* in Latin, and refers to the short Latin key for each species. It is not, in the philosophical sense, an essence; that is, it is not what constitutes the species. Instead it is merely a diagnostic key that permits identification in the field or at the workbench. Failure to recognize this has led to the essentialism story discussed in Wilkins 2009b.
23. See Williams, Ebach, and Nelson 2008.
24. Cain 1999.
25. Elkana 1984, Whewell 1840, on colligation. Mill 1974, 302; Bk III, chapter 2, §4.
26. Darwin 1859, 52.
27. Anonymous editorial, 1908.
28. Examples of Darwin interpreted as a species nominalist include Mayr, Ereshefsky and Stamos (Ereshefsky 2010a; Mayr 1969; Stamos 2003), but Wilkins argues that Darwin was always a species realist (except that Darwin did not accept there was a species rank, as Ereshefsky correctly notes, Wilkins 2009b).
29. Quine 1948. Hempel also influenced the issue widely, as outlined in Sandri 1969.
30. Braddon-Mitchell and Nola 2009; Quine 1953.
31. Hanson 1958.
32. Duhem 1954[1991].
33. This also has the unpleasant result that your ontological "world" is defined by all your theoretical commitments, so if you do not share all the theoretical commitments of another researcher, you are literally in a different world. In other words, if the ontology is constructed, nobody can disagree

rationally. There is simply no set of general observations that might arbitrate independently. This was an early criticism of Kuhnian "paradigms" Scheffler 1967.

34. van Fraassen 2008, 143f, 164.

35. Quine's claim that "[c]reatures inveterately wrong in their inductions have a pathetic, but praiseworthy, tendency to die before reproducing their kind" Quine 1969a, 126 suggests that he believed that there was something "natural" about these quality spaces.

36. Lewis 1970. "Ramseyfication" is the casting of a theory in some logical form, usually first order predicate logic, and deriving its ontology from the variables, based on the work of Frank Ramsey (Ramsey 1931 (1954)).

37. Faith and Cranston 1992; Kluge 2009; Kurt Lienau and DeSalle 2009; Rieppel 2003a; Siddall 2001.

38. *The Simpsons*, Episode: "On a Clear Day I Can't See My Sister" (2005).

39. Claridge, Dawah, and Wilson 1997; Ereshefsky 1991, 1992. This is discussed in Maclaurin and Sterelny 2008.

40. E.g., "deme" (Winsor 2000), "operational taxonomic unit" (Sneath and Sokal 1973; Sokal and Sneath 1963; cf. Blaxter et al. 2005), "evolutionarily significant unit" (Green 2005; Mayden and Wood 1995), and so forth. Linnaeus himself used the term "phalanx", given his martial and imperial language for the ranks of taxonomy (Stafleu 1971).

41. Simpson 1961, 7–11.

42. The term systematics also has two different meanings within biology, classification and, what those people do who call themselves systematists. For example, compare these two definitions of systematics:"Pertaining to, following, or arranged according to a system of classification; of or pertaining to classification, classificatory" [Oxford English Dictionary Online, accessed 12 June 2013].

"Systematics is the study of biological diversity and its origins. It focuses on understanding evolutionary relationships among organisms, species, higher taxa, or other biological entities such as genes, and the evolution of properties of taxa including intrinsic traits, ecological interactions, and geographic distributions. An important part of systematics is the development of methods for various aspects of phylogenetic inference and biological nomenclature/classification" [Instructions to Authors, Systematic Biology, *Journal of the Society of Systematic Biologists*: http://www.oxfordjournals.org/our_journals/sysbio/ for_authors/ms_preparation.html accessed 12 June 2013].

43. Ornduff 1969.

44. Schuh and Brower 2009, chapter 1.

45. Singh 2004, 2.

46. Mayden 1992, xviii. Kevin de Queiroz (for example de Queiroz and Gauthier 1992; de Queiroz 1988, 1992) treats taxonomy as being about naming and communication, and classification as synonymous with taxonomy. Systematics is, for him, the "ordering of entities into systems" (note 2 of his 1988). This is the exact inverse of Ornduff.

47. Stevens 1994, 11.

48. See Small 1989 for a discussion, and Lindley 1830 for the earliest use of "systematics" we can locate. On page 13 of the Preface, Lindley writes of the "synthetical principles of classification" and shortly refers to "systematic

Botanists"; it is clear that he intends it to refer to the process of naming and arranging groups.
49. Candolle 1819, 19.
50. Turrill 1935.
51. Gray 1879, 3.
52. Huxley 1940.
53. Turrill 1935, see also Turrill 1940, 1942.
54. Sneath and Sokal 1973; Sokal and Sneath 1963; see Winsor 2004.
55. Sarkar 1996.
56. Brundin 1966.
57. Hennig 1965; Brundin 1966, 1972a, 1972b, attacked by Darlington 1970.
58. Nelson 1973; Nelson and Platnick 1981; Patterson 1982b, 1988b.
59. Sober 1988.
60. Felsenstein 2004.
61. Personal communication in 2001. He has since published this in his book (Felsenstein 2004, 145): "A phylogenetic systematist and an evolutionary systematist may make very different classifications, while inferring much the same phylogeny. If it is the phylogeny that gets used by other biologists, their differences about how to classify may not be important. I have consequently announced that I have founded the fourth great school of classification, the It-Doesn't-Matter-Very-Much school."
    See also Franz 2005a. The site is http://evolution.genetics.washington.edu/phylip/software.html.
62. Devitt 2008.
63. Agassiz 1860.
64. Locke c1900, Bk III, chapter III, §15.
65. Abstract objects are objects not bounded by time and space, according to one view: Zalta 1988. Hence, any physical object, which is so bounded, is not a part of an abstract object.
66. The term "universal" is misleading. Aristotle used a portmanteau term *katholou*, which means "according to the whole" (*On Interpretation* 17a–17b). For him, a universal was a word, a predicate, that applied to two or more things. However, universals have inflated since to become general properties of the world as well as of words.
67. See Floridi 2004.
68. See Sedley 2007 for a history of this idea. The "creationism" here is of course not the modern kind. We may refer to people as "methodological creationists" – a term of Paul Griffiths', Griffiths 2009, 24 – if they assume that an object's classification does not appeal to a process explanation. This is a different meaning again.
69. Wittgenstein 1922.
70. This term was introduced by John Venn in 1866 but the idea traces back to John Stuart Mill's *A System of Logic* in 1843 (see Hacking 1991; Mill 1974, IV.vii.4, 720f; Venn 1866). An excellent overview of the issues is Magnus 2012.
71. Brigandt 2009, 79.
72. Boyd 1991.
73. E.g., Boyd 1999a; Griffiths 1999; Rieppel 2009; Schlichting and Pigliucci 1998; Wilson 1999a. Others are cited by Brigandt.
74. See Keller, Boyd, and Wheeler 2003.
75. Griffiths 1999.

76. Species can evolve more than once, from different stock, which has been dubbed the respeciation problem by Turner 2002. Moreover, many species comprise populational and haplotype lineages that are not monophyletic (Beltran et al. 2002).

77. Wilkins 2013a.

78. Aristotle's use of Greek terms to denote ideas is not, Wilkins believes, anything like the ordinary technical terminology of the scholastics, analytic philosophers, or modern scientists. Most of the time he simply grabs a common word and uses it more or less in the ordinary sense. An example here is *eidos*, which is translated as species in Latin traditions, but which just means something like the appearance or shape of a kind of thing. In this respect he is more like Wittgenstein or David Lewis (Lewis 1991) than, say, Carnap or Aquinas.

79. Aristotle is often used as a whipping post for various ideas that he has at best a tenuous connection; often he is attacked for the way medieval thinkers employed his views in science; for instance by Bertrand Russell (Russell 1950, 135). However, there has been a considerable renaissance in thinking of Aristotle as a good observer of the natural world, see Pellegrin 1986; Lennox 2001; Mayhew 2004.

80. Wilkins 2009b, 2010.

81. Whewell 1840, vol 2: 517–519.

82. Ably described by Stevens 1994.

83. Mill 1974, IV.vii.3, where he extensively quotes from Whewell's *History of the Inductive Sciences*, (Whewell 1837, vol II, 120–122). Mill saw himself in conflict with Whewell for political and moral reasons as well as for logical and methodological reasons (Snyder 2006, chapter 3).

84. Wilkins 2013a.

85. Wilkins 2009b, 129–159

86. Whewell 1837, Vol III, ch XVIII. 481f.

87. Hull 1981, 1984.

88. Hull and Wilkins 2005.

89. Bailey 1994.

90. The computer sciences have a tendency to invert the uses of terms they pillage from other disciplines. This causes all kinds of confusions. A recent example is the Ontology Project, in which ontology itself is treated as a kind of database.

91. Hennig 1975; Mayr 1974.

92. Chung 2003.

93. Winsor 2009; Winsor 2003, 2004, 2006b.

94. We use the term "real" without committing ourselves to either antirealism or realism of any kind. The term is used in the "good enough for government work" sense of everyday scientific life. It is enough to say that something is real if all observers in some investigative domain concur that it is, although one of us Wilkins is a structural realist of the kind described in Psillos 1999. We would not want to be accused, as Wilkins once was in conversation with Sahotra Sarkar, of having a "religion" with respect to scientific concepts and their referents.

95. Wilkins 2010.

96. Including by Wilkins 2009b, 2010, 2011.

97. As argued by Fitzhugh 2009.

3
Scientific Classification

Taxonomists may be described as producers, their productions being the classifications and names of plants. The non-taxonomists may be likened to consumers, the aforesaid classifications and names being the commodities which they consume. Now the characteristics of a commodity are of importance not only to the consumer but also to the producer, since, if the two are to continue in amicable trade relations, they must be satisfied with one another. It must be economic for the consumer to remain the customer of the producer. This the producer seeks to ensure by giving close attention to the standard and utility of his wares. [Ronald Good[1]]

Whenever scientists are presented with a system of classification, it is only a matter of time before they begin to ask whether there may be some underlying explanation for the pattern. [Eric Scerri[2]]

In this chapter we consider the sociological and phenomenal aspects of classification. The tribalism of taxonomy and systematics is discussed, leading to the tasks of classification, to order taxa and objects so that inferences can be made from them. Classing and ordering objects are distinct actions. We consider the iconographical representations of classification, and deflate "tree-thinking" somewhat. We note the influence on the thinking of classifiers of the ontological fallacy, believing that because we have given a name to a group we think we see, that group must exist. Finally, we discuss names and nomenclature.

The Sharks and the Jets

A problem with the dynamic social conception of science is that it makes science vulnerable to the general problems of all social behaviors; that is, it makes science political and sociological. By this we refer to the politics of the sciences themselves, rather than the secular partisan politics of the local national system, although they, too, can affect science dramatically. However, here we mean that science forms into rival research programs, laboratories, projects, national academies, universities, and so forth.[3] In short, science becomes a performance of Leonard Bernstein's *West Side Story*, and rival groups act out the Sharks and the Jets, only without the [overt] jazz ballet.

Classificatory activities and practices exhibit an extreme tribalism in many sciences. It appears that how one names and groups the objects of study acts as a surrogate for the political affiliations and alliances that the social structure of the science concerned has evolved. This is not, in itself, a criticism, but simply a fact of the nature of science.

There are several issues in play. One is, as we have said, a matter of professional strategy, and this both depends upon what the professional scientist expects will pay off, and upon the relations that scientist has with other professionals. Another is the cognitive "cost" of learning a taxonomy in order to employ it, to communicate with others, and to be able to educate nascent professionals. A taxonomy plays a substantial role in the practice of science, and so we should expect that research groups will tend to coordinate on one or a close cluster of solutions, by convention.[4]

However, this results in strong tribalism, in which the universalistic tendencies of science and the academy in general are dampened in favor of a kind of conceptual nepotism. It would be naive to think this is rare in the academy. In some traditions, only those views that are congruent with the opinions of authorities can be published or taught; a holdover of the kind of authoritarianism to be found in the older European traditions. We call this German Authority Syndrome, based upon the behavior of nineteenth century German-speaking systematists, although it is unfair to burden only the German tradition with it, since it is widely observed in systematics generally (and in the past, in European systematics especially). The more or less subjective decisions of an Authority become mandated and regulatory. Gary Nelson once noted that the original task of one school of classification, phenetics was to oppose the practices, said to be "traditional," of grouping and ranking with the arbitrariness of a monarch's drawing boundaries and conferring titles.[5]

This applies at both the object level of particular taxonomic decisions, and at the metalevel of taxonomic philosophy and protocols. This subjectivism has ever been a problem of biological systematics. By analogy it is likely to be a problem of systematics among other disciplines as well, including the new fields in biology where systematics are applied (such as molecular systematics and phylogenomics, along with many other fields that end in -omics).

The evidence for the tribalism of biological systematics is anecdotal, although David Hull noted that Sten Lindroth once called systematics "the most lovable of the sciences".[6] We suspect he meant the concepts of systematics rather than the sociology of the discipline. Hull himself studied the acceptance and rejection of differing schools of taxonomic thought in the journal *Systematic Zoology*, and found no consistent bias by editor, but the decisions by authors over which journal to submit to are based on a shared disciplinary knowledge of which journal, and which editor, accepts papers on which systematic techniques and philosophy.[7] A wider study than Hull's would be needed to show the actual sociological relations and how they correlate with systematic philosophy and approach, as anecdotal stories are often biased and sometimes actionable in court. However, such battles are played out in the book reviews and critical papers.

This has a regrettable effect on debate. Tests for orthodox opinion are imposed based on the use of the "correct" terminology, techniques and underlying philosophical justifications, with many refusing to even consider contrary views. Those who are thought to be heterodox are attacked, often personally, and as a result of the chilling effect either stop discussing these issues or become resentful and defensive. For example, pattern cladists have been called "creationists" and "antievolutionists" by orthodox cladists; and yet, most of the points made by the pattern cladists about the difficulty in retrieving history, the problems of homoplasy, and recently objections to the simplistic use of molecular data, either have been taken into mainstream opinion or are in the process of being adopted.[8] Often, all that remains are the labels. Likewise, even as advocates for numerical, evolutionary or Linnaean systematics deride the cladists, they are quietly adopting monophyletic groups in their taxonomies, and cladists are employing the algorithms of phenetics, and so on. To an outside observer, systematics looks very much like the various Judaean People's Fronts from Monty Python's *Life of Brian*:

Francis: What ever happened to the Popular Front, Reg?
Reg: He's over there.
P.J.F: Splitter![9]

Stephen Toulmin once noted that what separated the catastrophists and the uniformitarians at the beginnings of the nineteenth century had by the latter half become a disagreement mostly about terms. The catastrophes had become smaller and the gradual changes of geology became more episodic until little of substance divided the two schools but the terms they used.[10] Likewise, many of the supposed philosophical differences between schools of thought in systematics in biology resolve eventually to political influence and alliances. This is not to say that there are no substantial differences, for of course there are, but only to note that the battles of the past are obscuring those real differences and making a general overview of the nature of natural classification hard to outline. It is our task in this book to look at the issues, both the differences and the shared consensus under various terminologies. To that end, we will later outline a general scheme of classificatory terms under the neutral heading of Radistics, which relates to the roots of the classificatory task in cognition.

We need to take all classificatory activities in science seriously. Indeed, we need to take all cognitive activities seriously, but as this is a book about classification, we will limit our scope of discussion. If scientists are doing classification by way of similarities, or by homologies, or by causal processes, or by essential properties, or merely by convention, then all these are important in science, no matter what the underlying philosophical foundations may be (and often, these are not so much foundations as they are post hoc justifications for what is already being done).

Historical considerations

At the beginning of the nineteenth century, around the time Charles Darwin headed off on his world tour (Rio! Sydney! Cape Town!), taxonomists were exercising themselves greatly over what was a "natural classification" in natural history, roughly in biology and geology. The shared view was that, as the system of Linnaeus was artificial, relying as it did solely on the sexual organs of plants, or a few characters in animals, the classification that resulted was at best only roughly telling us what the nature of things were.

Competing systems included that of Jussieu and Adanson, who used many characters to draw up their tables of classifications.[11] But the classifications themselves were regarded as being statements about the natural order. The "method" used was independent of that fact. Classifications were attempts to work out the order of nature. But there was no real theory involved; these classifications were made on the basis of empirical investigation. "Affinities" were identified on the basis of the overall characters of organisms. There simply was no theory. Just as species had been identified long before even the rules of heredity were understood, so too classifications were done before the science had explained them. Indeed, the classifications themselves triggered the development of theory, most notably that of Darwin's theory of common descent.[12]

How is it that the classifications that early naturalists developed, and today which remain largely relevant, if not still entirely valid, was done without theory, if the Quinean account is correct? The answer, it seems, is that the account is not entirely correct. It does not exhaust the possibilities of the dynamic of scientific investigation. Induction itself has had a revival among the "new experimentalist" philosophical movement led by Hacking,[13] but classification has had no such revival. It is time it did. How that is to be undertaken is the topic of this chapter.

The tasks of classification

What is it that classifications are intended to achieve in science? In library catalogues and other conventional classifications, convenience is the sole criterion.[14] It is enough that the items classified can be found, stored, retrieved, indexed and referred to. This aspect is present also in every natural classification; for scientists must talk to each other about the same items, store and retrieve them in museums and so forth, and be able to process metadata about them. The notion of "metadata", which is taken here from markup languages for digital content management systems, is crucial, in that the metadata is itself a classification, but it is a classification about the objects, not a classification of the objects. Conventional classifications are classifications of the tasks that the users of the objects being categorized wish their system to undertake.

Natural classifications, on the other hand, are classifications that begin and end in the properties of the objects. While they must be useable by those specialists who need them, the primary task of a natural classification is to arrange the objects in terms of the properties of the objects themselves, and not the properties of the investigators of those objects. A natural classification says something about the phenomena of a domain, not about the scientists who study it, and an unusable natural classification may be natural if the domain being classified really is so complex it is incomprehensible by even a specialist (for example, if it was done by computer).

We must revisit the distinction between taxonomy and systematics again. Taxonomy is the professional practice of naming taxa, storing, retrieving and arranging data, and of teaching field workers and students the objects of the subject. It involves propaedeutic tractability. A subject cannot be learned which is unlearnable, and so we must organize our material so that it can be learned. This is Linnaeus' task: to set up a system that is useful for communication and learning. However, while cognitive psychology matters, it is not a matter of fact about the things being investigated. Systematics is about organizing the taxa into ordered arrangements. It also has a conventional element, but the focus, in biology and other fields, is largely a matter of discovering the "natural" arrangement, in order to license further inferences about the taxa. It is crucial to distinguish between what is convenient to science (conventionalism) and what is informative (naturalism). Failing to do this is a common mistake in discussions of classification and very largely underlies the shift at the end of the nineteenth century from the belief that classifications, especially in biology, were informative to the belief that they were simply an act of conventional librarianship.

The process of classifying

A classification involves two "moments" or activities: one is the setting of things into equivalence classes, or types, which we shall call classing (included in taxonomy). We shall later ask in what ways this is done, but that it is done is the first crucial element of a classification. The second is the ordering of the classes (included in systematics).

Classes of this kind are variously characterized. Terms like class, type, kind, group and others are employed and given differing interpretations. One widely held view is that classes and kinds must be defined by necessary-and-sufficient-properties (either semantic or physical), while groups need not be.[15] Types are regarded as, in the words of one reviewer of a paper, "a throwback to a two hundred year old metaphysics", and the resultant classification is called, derogatorily, a typology.[16] So-called typological thinking is regarded as a profoundly un-Darwinian and fallacious approach to classification, and the alternative, correct, view

Mayr called population thinking, in which the equivalence classes were groups of organisms that varied on distribution curves. We will be agnostic about this, for it seems to us that all classification involves types, that types are less malign than they have typically been painted, and indeed that Darwin, among others of his time, was himself a typologist, as all systematists since have been, including Mayr. This is because types have been misrepresented, as we argued in the previous chapter. Moreover, a population that has a single normal distribution for its traits has a mesotype in any event, and when it has a plurality of distributions, it must have some consensus of mesotypes, or it would not be recognized as a population.

Once the classes have been assembled, either as types or populations, they must be arranged with respect to each other. The reason for this varies depending on the classification's role and underpinning. Since we are concerned only with natural classification, we will ignore the communicative aspects in favor of the inferential aspects of a classification. In short, the basic purpose of arranging taxa is to license inferences of a general nature on that basis.[17] It is often held that classification is an early and naive task of a science, and the more Baconian view that Whewell promoted suggests that there are predetermined steps to go through (as developed in Chapter 2). First one should recognize the classes, and then one should make the taxonomic arrangements (Figure 3.1A). On the more recent theory-dependent view of classification, however, this is inverted. One first orders the classes and then one identifies the elements or members of the class (Figure 3.1B). A

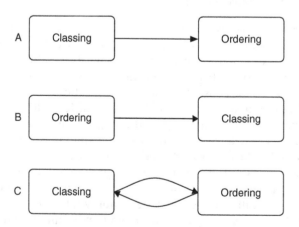

Figure 3.1   Sequences or moments of the task of classifying natural taxa

more realistic view, however, which matches both the iterative nature of science and the history of scientific development, is that there is a "reciprocal illumination" (a term we take from Hennig) between each moment. One may have a better grasp of what counts as the class and membership of it in one case, and the problem being presented is to find the most empirically adequate ordering, or one may have a good sense of the arrangement of the taxa but find it hard to work out how to either identify members, or characterize the taxa in general. In such cases it is best to attempt to incrementally refine one's classification in both directions (Figure 3.1C).

The general criteria for classing objects into a single equivalence class are many. One may do this by shared properties, by functional equivalence, and by clustering (in the context of biology, clades, grades and clouds, respectively). The properties employed may be physical properties, theoretical properties or semantic properties (which we exclude from natural classification unless the subject is semantics). Since classing objects is an epistemic activity, for this to be natural classification, it must be that the classing agrees with states of affairs in the extra-epistemic world. The classes must represent the world in the right way. This, of course, presumes there is a right way to represent them. If the world has structure independently of the way we represent it, and it should be a basic commitment in the sciences that there is, then there is at least one right and a number of wrong ways to represent the world. There need not be a single right way, as there may be many tasks for these representations to play, each of which has its own priorities, but given those tasks and priorities, one can do it rightly or wrongly if the structure is accessible.

The critical issue about placing objects into classes is what basis might be used, and this depends upon a relevant identity between the objects classed. We shall call these objects specimens, which is a general term that has several meanings in specific sciences such as biology or geology.[18] To class specimens, they must be "the same" in some respect. To ensure that a class is natural, that respect must involve facts about the specimens (as opposed to facts about the classifier or classification system). This means that a class has to have natural properties that other classes do not have.

It is widely agreed that natural classes need not have unique and universally shared sets of properties in a science. The argument has been most extensively made in the case of biological species:[19] not all members of a species need have a set of traits or properties (including molecular genetic properties) that all and only members of the species have. Some species may, in contingent fact, have such properties, but while it is nice when they do for philosophers, it is not required of all members of all species that this be the case.

Some have suggested that all members of a species have extrinsic shared properties – usually genealogical or some functional shared similarity. The genealogical account, exemplified by Griffiths' "historical essence" account,[20] supposes that all and only members of species have a shared common ancestor; however this is mitigated if not refuted by the existence of species in which the last common ancestor of all members of a species is also shared by members of other species, either through repeated speciation ("respeciation"[21]), hybridism, or incomplete coalescence of genetic lineages across the speciation event. Moreover, one may imagine an extinct species being resurrected, say through clonal techniques from a sample, in which its parent is a member of another species, and so on. We would still say that a resurrected woolly mammoth was a member of *Mammuthus primigenius* (although some may not).

The functional shared similarity account is best represented by the homeostatic property cluster account of Boyd and colleagues.[22] According to this, all specimens of a species are part of their class because they contribute to and are constrained by a shared set of properties that act causally to maintain the group as a coherent functional object. However, this involves the claim that any group in which there are several such ensembles of homeostatic properties is not a single group. In what biologists call "polymorphic" groups, this may not necessarily apply. But it appears to be on the right track, in our view. We will expand upon this.

It may help if we can say what, formally speaking, makes a class a class. Contrary to the assumptions of species individualists such as Hull and Ghiselin, it is not necessarily the case that a class by definition is a group in which there is a set of necessary-and-sufficient-properties all specimens possess; at least, not in biology and other sciences. It may very well be the case in logic, or some logical accounts. Set theory need not be interpreted as requiring these intensional properties, however. A class in this sense can be a group of objects that happens to be adjacent, or which are circumscribed by extrinsic properties (such as "the class of all objects in Bertrand Russell's left trousers pocket on January 20, 1914 at 12 noon GMT"). Objections to classes on the grounds they involve universals or essences are misplaced. However, if one prefers to think of classes this way, substitute "set" or "bag" or some other group-container as we use the term.

For something to be a class, its members must be "the same" in some manner. Another way to say this is to invoke Leibniz's Law: all members of a class are indiscernibly different. If you take one specimen of the class and arbitrarily replace it with another, it remains the same class, and the specimen has all the ordinary properties of the class members that the other one had. This implies that in a natural classification, with which we are here concerned, the specimens must be indiscernibly different with respect to some properties that are physical. Now in the case of Russell's pocket, I can put my hand on the house keys or the pipe, and while they are different in many ways, they are not different as objects in Russell's pocket. The porcelain elephant on the table, which will not fit and is not in Russell's pocket, is discernibly different with respect to the property used for inclusion (literally) in the class. So we can exclude it, on the basis of physical properties.

Of course this trivial and eclectic example is not very interesting to a science. Sciences want specimens to group into a class on the basis of interesting (in the science) inclusion criteria. But we may not know, a priori, what those criteria are. In fact, when a science or investigation begins, we do not know. How do we group under uncertainty? The answer to that is not simple. It is usually presumed in the philosophical literature that we group naively through similarity, but we shall argue later that similarity is not sufficient in many cases, and that there are too many similarities that we might use. Instead, we shall argue that grouping is done through identities. That is, we put things into a class when they are all the same. Not the same in every respect, for outside physics and the universal sciences, entities and specimens will never be the same in all respects, but the same in the respects that this science or this investigation requires.

The problem lies in the fact that we cannot know ahead of time what properties these should be. Do the number of hairs on the ear of a dog matter for its classing into *Canis lupus*? Maybe it is obvious that they do not, but why not? After all, the number of hair like structures on the genitalia of, say, a beetle might very well causally maintain that species. Until we know what the causal roles of the traits are, we cannot tell if the characters matter for classing. However, until we have the class in hand, we cannot find out what the relevant characters might be. The conundrum is resolved by realizing that at no stage in an act of classification do we approach the subject tabula rasa. We always know something about the domain (or else we'd not be investigating that domain; we might not even recognize its existence). We have prior information used for inductive generalizations.

The basic operation of ordering is this: A is more closely related to B than it is to C. Ordering taxa means to assert a relationship of more or less close proximity between at a minimum, three taxa. Two taxa can have no relation that is natural, for there are infinitely or at the least indefinitely many ways in which two objects may be related, a point first made, we think, by Locke.[23] The pattern of three taxa, on the other hand, forms a relationship, and all classificatory relationships are therefore composed of three-taxon patterns. Larger classifications are composed of assemblages of three-taxon patterns. (This is a slightly different point to the "three-item analysis" of biological systematics; it is purely a logical point rather than a methodological one. In this book we are not concerned with methodologies or implementations. Scientists will use whatever methods they find give the best results, and if that turns out to be "parsimony", Bayes, or likelihood, or some arcane statistical procedure, so be it. However the arrangements are arrived at, the logic of the ordering is independent.)

What is it to say that something is related to something else? After Locke, this was often held to be a purely conceptual or verbal matter, and in Whitehead and Russell's work, relations were logical predicates.[24] But in natural classification, if it is possible, we must have real relations, no matter how we might interpret "real". That is, if there are relations in a natural classification, they must be as real as anything else can be in the natural sciences. Locke himself identified one kind of real (as opposed to nominal or semantic-logical) relation: genealogical relations.[25]

A relation in the natural world is some etiology. That is, a relation is real, causal, and important. There are many possibly important real and causal relations between objects (such as the gravitational influence on molecules in my ear by *Eta Arcturus*) but the truly important ones are those that most contribute to things being what they are (the classes they are members of) and how they are arranged relative to other such entities. Unfortunately, as is often noted, the natural world fails to come with reliable tags identifying these classes and arrangements, and so we construct the classifications under considerable uncertainty. We cannot merely infer deductively from known classes, because the classes themselves must be rebuilt at sea like Neurath's Raft. It is a common error among systematists (especially those who take Popper at his word) to think that relations are logical and deductive, when in fact all "deduction" about them relies upon prior induction just to construct the classes that bear names and properties.

In biology, relations are the outcome of reproductive lineages. All accept this truism. But there is ambiguity when talking about relations: are they the reproductive lineages (which are usually not known and must be inferred from observations) or are they the formal arrangements of taxa (which are abstract constructs)? In fact they are both, and it would help communication if we distinguished between real relations and relationship statements, which summarize the affinities of the taxa. A cladogram, for example, is a relationship statement, while an evolutionary tree (if known and true) would represent real linear relations.

In other sciences, where etiology is not constrained by shared history (for example, in geology where rocks can be formed the same way independently), relations are purely causal; but even here the initial conditions mean that one may identify a specimen in a class as being from a particular locale or period based on contingent constituents. For example, isotope ratios make identifying radioactive material sources relatively simple, as many thriller writers have realized.[26] "Blood diamonds" are identifiable based on impurities unique to the locale from which they were mined.[27] A relation here is some etiology that is shared to members of the class, which makes or contributes to the ensemble of properties that make a specimen a member of the class. This would be a tautology if not for the fact that the construction of the class itself is, in the first instance, an iterative process of refinement of the class based on the success of inferences made from it. This is also very like the process Hennig called reciprocal illumination in another context.[28]

Iconography of classification

The most contentious aspect of modern systematics lies in the iconography, or illustrative metaphors, used. "Tree-thinking" is only one aspect of this. In the period after the foundation of the Modern Synthesis, so-called, evolutionary trees were frequently drawn, each of which had problems in the implications and connotations they carried, especially in popular thought.[29] The history of classifications does not support the idea that tree-thinking was either novel or particularly significant. The earliest use of a tree metaphor in systematics is in Peter Simon Pallas' 1766 book *Elenchus zoophytorum*, in which he describes the relations of plants and animals as branches on a two-trunk tree. Pallas was rejecting Charles Bonnet's Échelle *des êtres naturels*, a form of the Great Chain.[30] In the period after Linnaeus became familiar, it was common to list taxa with indented lists, which became "tables", usually with large braces. These were also used in logical texts, such as George Bentham's early text on logic discussing Whately's *Elements*.[31] The earliest tree diagram in natural history is Augustin Augier's 1801 diagram.[32] Apart from Ernst Haeckel's "mighty oaks", a diagrammatic system of trees did not develop until late in the nineteenth century; for example, William A. Herdman's "phylogenetic table" (Figure 3.2),[33] in which grades and clades are mixed together.

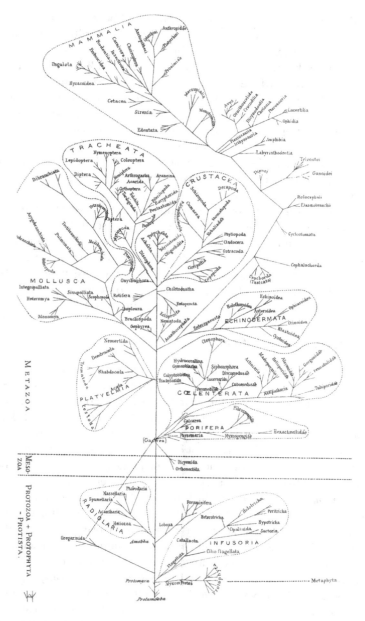

Figure 3.2   Herdman's 1885 "Phylogenetic Table" of animals, showing the topology of relationships, but with the vertical dimension representing grades, shown with dotted horizontal lines: "The lowest organisms are placed at the foot of the Table, the highest at the top." Grades are clearly based on something like the Great Chain here. However, note that he permits "retrograde" evolution for the "Mesozoa"

Early tree diagrams in systematics do not represent history, of course (the earliest that does is Heinrich Georg Bronn's in 1858, or perhaps Edward Hitchcock's in 1840), although in linguistics there are some early history trees.[34] They instead represent the hierarchical nature of the classifications, and this requirement for classifications was well elaborated by Linnaeus, and indeed described (for "predicates", or words) by Aristotle in the *Posterior Analytics* (74b). The invention by Euler of what are now called "Venn diagrams" (Euler invented the use of ellipses to indicate the scope of classes and their intersection; Venn added only the shading of intersects) was another graphical technique to indicate this: a proper set, as we would now call it, included non-intersecting subsets. This is clear in Hennig's own famous diagram inventing cladograms (Figure 3.3).

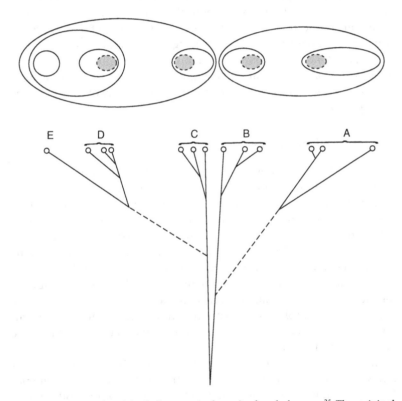

Figure 3.3 Hennig's original diagram (redrawn) of a cladogram.[35] The original caption reads "Morphological divergence and phylogenetic relationship". Venn diagrams are added above it to show that the topology of the tree is equivalent to a formal relationship statement – the grayed dotted circles Hennig did not label. The relationship statement would be {{{E, D}, C}, {B, A}}, although this does not include the unlabelled taxa

Hennig's diagram, like Herdman's, attempts to represent both degree of difference (in Hennig's diagram, morphological divergence, and in Herdman's, degree of evolution) and relationship. Later cladograms do not attempt to model degree of difference, although Mayr clearly thought that was a function of cladograms like the early Hennig. As we shall see in the section "Hybrid classifications" in Chapter 5, this conflates the representative tasks of two kinds of classification.

How we represent the relationships is largely independent of what they represent. In biological classification, a relationship is something that we reconstruct from our knowledge of the organisms in that group. Presumably it represents some causal chain, and so a cladogram or other format (indented list, relationship statement, brace diagram, tree diagram, Venn diagram, etc.) is an attempt to indicate the outcome of that process. It is not, however, identical to a statement of that process, since outcomes can be multiply realized. The history of a given group of organisms might be the same topology as a relationship statement, but it might not (and, purely in terms of the number of topologies of historical processes that could produce the observed relationships, almost certainly is not). We will revisit the iconography of classifications in Chapter 7.

The ontological fallacy

One of the enduring mistakes made in science and philosophy is to confuse how things seem with how they are. In biology, the conjunction "pattern and process" has been around for decades, at least since 1947 in ecology, when Alex Watt used it as a title for an essay on plant communities.[36] In 1967, Terrell Hamilton published his *Process and Pattern in Evolution*,[37] and the phrase took off from there. But patterns are just relationships that are salient to the observer, while the processes account for the patterns. In the ontology of biology, it is common for theorists and critics alike to argue that an observed pattern of some kind just is the ontology – that communities are real, that species are an organizational rank in biology, that groups are formed by an "adaptive radiation", and so on.

The mistake of this conflation of epistemology with ontology (and not always a well-founded epistemology – often it is merely a phenomenological and subjective impression held by some authority or other) is an old one, but as a fallacy it was not named until Alfred North Whitehead called it "the fallacy of misplaced concreteness".[38] Later, Herbert Marcuse called it the "reification fallacy".[39] We call it the ontological fallacy. It is the error of taking a noun to require a concrete referent, and in this case it is the error of thinking that because one sees a pattern, there must be a process that the pattern answers to. However, as pareidola indicates, the pattern recognition systems our brains are capable of identifying patterns that aren't there (or more exactly, identifying patterns that do not have any other homogenous underlying natural process than the act of pattern recognition itself). It should not need to be said that this is a mistake, but it is a subtle one that occurs frequently in the literature. It is not a foundation for an ontology, and such observer bias is ideally eliminated from scientific research (an ideal honored often more in the breach than the observance).

Take, for an example, species. Many biologists identify species as a salient grouping of organisms, and the salience is based on the sharing of a group of traits, or typical features. The majority of reflective biologists know that these marks of identification or diagnosis are not the causal processes that make the groupings in nature; that is more to do with an interaction between shared developmental systems including genomes, cell types, and behaviors, and the environments and ecologies in which they exist. This is sometimes referred to as the "interactionist consensus".[40] It is clear that if one thinks that species have different organizational structures in different parts of the evolutionary tree (plant species and mammal species are quite different, and bacterial species even more so),[41] there can be no such thing as "the" organizational rank of being a species. But we have encountered, both personally and in the literature, biologists who sincerely believe that because there is a typical organizational structure for species of, say, snakes or birds or ants, that this is the rank and structure for all species; a classic example of putting pattern before process.

Systematics is the hunting ground for this mistake par excellence. Often people think that classification is just the process of assigning names, and that once you have a name you have a thing that is named. The use of clustering algorithms to identify species in molecular data is as much the ontological fallacy as phenetics had been when it was used for morphological data – what such algorithms tell you is that the criteria for grouping work in such and such a way on that data set. What clustering does not tell you is that there is a significant and natural process underlying it that makes the pattern informative. The current underlying justification for phylogenetic classification is that the patterns it uncovers are supposed to be the outcome of historical processes of descent with modification, or as we call it, common descent. The problem with other methods of grouping organisms is that at best they merely mark one process (the one on which the clustering is based), and at worst they reflect only the choice of principal components, which are epistemically relative.

Ontologies are supposed to mark the classes or kinds of things that exist in a domain or under a theory. However, like any discipline that deals with the empirical world, all we have access to are the patterns, so over time science has developed many methodologies to attempt to ensure that only the patterns that are actually informative about the world are the ones that get used. There is no magic "noetic ray" that will do this, as Putnam once noted,[42] so we are stuck with the patterns in our data sets from which we seek to reconstruct the processes. This is as true in physics as in biology, and it lends itself to the ontological fallacy in science. If we have patterns in our data, and we use them to figure out the processes, it is a short leap to thinking that the patterns are the processes, even if on reflection or under challenge, a scientist will say "Well of course the data isn't the things themselves!" As Joel Cracraft once said of species "...somehow a species definition must be inclusive of an ontology and an epistemology".[43] Well, there's the trick.

As patterns do not automatically give us the processes, they do not help us with our ontology, however they may play out in our epistemology. Even the choice of what to measure and how to measure it is already presumptive of an ontology, and really we ought to have some foundation for that before we get going, which implies that we already have our ontology before we apply our epistemology, and this can't be right. Where does the ontology come from? If not from a prior theory (as a variable in the Quinean sense), it cannot come in Aristotelian fashion by refining definitions or intuitions – that project of science by definition died in the sixteenth century, although the body still lingers rotting. The answer is that we evolve our ontologies, so what we apply at t is the best outcome of what we obtained at $t - n$. In doing so, we have always to be on our guard against the ontological fallacy.

Names and nomenclature

One of the main focuses in the literature, especially in biology, regarding classification is the problem of nomenclature, of names. Many treat classification as being all about names, an error that is akin to mistaking not the map for the territory, but the names on the map for the territory. Recognizing this, many biologists and philosophers think of classification as just about names, subject to the philosophical problems of names as definite descriptions, with the associated problems of opacity of reference, and so on. Treating names in classification as merely a matter of semantic convenience has led many to think that classification is not something terribly deep in the natural sciences.[44]

Naming has a conventional element, of course. It does not matter much if we call a species by a Latin name or a German name, so long as the name is unambiguous, and used by everyone. The reason why Linnaeus' scheme was so successful is that he took the confusion of folk names in various languages, and the definite description names in prior work (in which the "name" of a taxon was a description of key characters up to a dozen words long, hardly useful) and replaced them with a simple two-part name, of genus (the wider group) and species (the most restricted group). This convention meant that communication was now simpler, and ambiguities of reference no longer a major problem, although of course that reference was now the focus, both in terms of taxonomic lumping and splitting of names, and in terms of what it was that the names did refer to.

Systematists are often very attached to the naming schemes for their domains. There are a number of reasons for this. One, the most obvious and important, is the amount of investment of time and resources that a naming scheme requires. It takes effort to find, describe, and name a taxon, and then for students and other specialists to learn it and the scheme within which it resides. Changes to this scheme are expensive, and so there had better be a good reason to make them. Nevertheless, there is a constant rate of turn-over in names and groups in traditional taxonomy. In the pinnipeds [seals and their relatives], for example, around 40 percent of taxa have been renamed, collapsed into synonyms, or split into different groups, at all ranks since Linnaeus first described them.[45] Why does this happen?

A naive empiricist might think that it is driven largely or entirely by new data, better descriptions and specimens, and so forth, but there are many other reasons, often social, why a taxonomy might change, and they are quite often the same reason why a taxonomy might also remain unchanged, or be defended:

1. National or ethnic preference and status. Naming rights in the imperial period were a matter of status and political power. People who were trained up in a given system tend to prefer their own scheme over those of competing nations. While Britain adopted the Linnaean system early in the nineteenth century, largely under the influence of James Edward Smith, at first the French rejected it in favor of the "total evidence" approach of Jussieu, and this tension persisted until

after the middle of the nineteenth century. Many German authors rejected both in favor of a largely morphological scheme based on Goethean or *Naturphilosophische* principles.

2. Disciplinary history. For example, the anatomical nomenclature that developed within medicine for bones and musculature is largely independent of the nomenclature developed for animal anatomy in veterinary science and biology. This makes cross-comparisons rather difficult at times.[46]

3. Politics in science. As naming rights are a measure of standing in some sciences – whoever names the taxon first is going to be cited extensively – attempts to change or defend a taxonomy will depend on whose interests are being advanced. This applies at the individual level, but also at the level of research groups and programs, schools, and of course professional institutions. Competing interests drive much of the debates and conflicts in nomenclature.

4. Priority. This is entirely legitimate, in that the work of an individual should result in their conceptual inclusive fitness increasing. This is a term taken from David Hull's evolutionary account of science,[47] and refers to the amount of cited use that a scientist's work receives. It is, in effect, imputed credit. Priority of discovery is supplemented by priority of naming; whose terminology is used is an indicator of the standing a scientist has in their discipline. Attempts to take away naming rights can rest on priority disputes, both in terms of claiming priority for someone else, but also as attempts to deprecate the work of a rival within a competing group or tradition.

5. Protectionists. Also known as "Lumpers" and "Splitters", taxonomists have a tendency to increase or decrease the number of given taxa based on a form of protectionism. Taxonomists who are experts are those that intellectually "own" a given taxon; say, a genus or family. With an increase in the number of people entering a field, the number of species will automatically increase, while the number of available specimens remains constant. A higher number of "new" discoveries per year, for example, creates a larger taxonomic field with further opportunities for growth, as well as establishing an inflated career. Splitters are responsible for an inflation in taxonomic divisions such as "sub-" and "supra-" taxa. When the field contains fewer specialists, a new generation of taxonomists revise existing groups and "lump" taxa, generally removing the "sub-" and "supra-" groups. This also includes removing monotypic taxa, such as subgenera or subfamilies that contain a single lower taxon. Lumpers may also boost their own career by specializing at higher taxonomic levels.

Friction between lumpers and splitters occurs when the generation "turn-over" overlaps, forming two separate classifications that are promoted simultaneously.

Names are not, in themselves, natural facts. If what they denote are natural objects, then nomenclature is critical in classification; otherwise it really is a conventional matter. So the issue is not whether the names are right, the issue is whether or not they unambiguously denote facts about the world. And here another core problem arises. As Leibniz wrote, paraphrasing Locke (Philolethes is Locke, Theophilus is Leibniz)

> PHIL. §25. Languages were established before sciences, and things were put into species by ignorant and illiterate people.
>
> THEO. This is true, but the people who study a subject-matter correct popular notions. Assayers have found precise methods for identifying and separating metals, botanists have marvelously extended our knowledge of plants, and experiments have been made on insects that have given us new routes into the knowledge of animals. However, we are still far short of halfway along our journey.[48]

Vernacular terms like "ape", "bird" and "tree" get a major revision by technical science. Some terms, like "mountain" or "stone" can be dissolved either into many technical terms, or spread ambiguously across terms of art in ways that make them scientifically meaningless. When the claim is made in the popular press that "birds are dinosaurs" or "humans are apes", there is a vernacular sense in which this is simply false. Every child knows that dinosaurs are flightless things with no fur that lived more than 65 million years ago (unless they are well educated into the arcane debates in paleontology; never underestimate a motivated ten year old), so how can birds be dinosaurs? The answer is, of course, that modern phylogenies (classifications based on shared traits that are thought to be the result of evolutionary history) place theropods (a type of dinosaur), squarely inside the group Aves, the taxonomic name for birds. and so by the rules of technical biological nomenclature birds (Aves) "are" (fall inside) dinosaurs (Dinosauria). However, the rules of nomenclature are quite strict and as Aves is the older term[49] it subsumes Dinosauria.[50] But nomenclature is only one part of the argument.

Classification is governed by strict naming rules, and so it should be, otherwise we will end up with a highly unstable classification in which we are unable to refer to same taxonomic groups (which is why *Hibbertia* can refer to both a trilobite and a herbaceous shrub). A form of logic also takes precedence. What if, for arguments' sake, a name also refers exclusively to a list of diagnosed characteristics? In this sense, Aves or birds are a very specifically defined group. We know what birds are, but dinosaurs are less known. For instance, the diagnosis for dinosaurs needs to be constantly altered in order to keep up with newer discoveries. The original description of Dinosauria did not include a vast majority of the characters that are now included in the group. Aves (or birds) on the other hand can still be diagnosed by the characters originally described to them. In other words, taxonomists know exactly what a bird is. However, they are uncertain as to what characters diagnose all dinosaurs. The same is true for any well-defined, usually natural, group when compared with a poorly-defined, usually artificial, group. Vertebrates are animals with backbones. Invertebrates do not have backbones. So, one may ask, what defines invertebrates? The answer is clearly what they are not, namely not-vertebrates. The same is true for prokaryotes (non-eukaryotes) and sauropods (non-avian dinosaurs). Relying on a non-taxon is in breach of a rigorous scientific taxonomy, as Figure 3.4 illustrates.

In other words, a natural group validates the existence of an artificial group after a taxon has been discovered to be artificial. Dinosaurs alone are artificial, but with the addition of birds (in hindsight) they become natural. In this sense Birds + Dinosaurs + some mammals qualify as reptiles (reptiles alone are artificial). Reptiles + mammals + birds qualify as fishes, as fishes alone are artificial. Reptiles + mammals + birds +

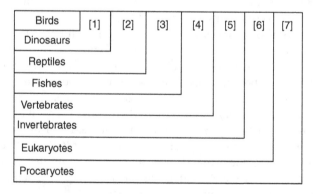

Figure 3.4 Organisms and non-organisms: [1] Non-Avian Dinosaur (e.g., Sauropod); [2] Non-Dinosaurian Reptile (e.g., Crocodile); [3] Non-Reptilian Fish (e.g., Frog/Goby); [4] Non- Fish Vertebrate (e.g., Hagfish); [5] Non-Vertebrate Invertebrate (e.g., Coral); [6] Non-Invertebrate Eukaryote (e.g., Avocado); [7] Non-Eukaryote Prokaryote (e.g., Bacteria).

fishes are vertebrates as they all contain vertebrates, but this was known before we discovered the naturalness of vertebrates. Where the dinosaurs are birds dispute derails (other than in nomenclature), is in how we read trees. Classification works its way down a tree.[51] A finch is more closely related to a species of theropod than it is to a species of sauropod. Phylogenies, however are read up a tree. So some Archosaurs gave rise to Dinosaurs, some of which gave rise to birds. This simple misunderstanding arises from how taxonomists approach classification.

Firstly there is, and mostly always has been, a quasi-scientific or folk taxonomy. This is then refined using strict nomenclatural rules (taxonomy). These taxa and their characteristics are then tested for naturalness (systematics). The resulting natural groups are interpreted phylogeneticaly by evolutionary biologists to determine ancestor–descendant relationships. In other words, taxonomy precedes systematics which precedes phylogenetics. This system changed during the mid twentieth century, when taxonomic systematics and phylogenetics were combined in a single methodology.[52] Now systematists determine taxonomies solely on phylogenetic interpretations. The resulting trees show that birds are closely related to dinosaurs, but phylogeneticistsinterpret this to mean they are derived from dinosaurs, therefore they must be dinosaurs. This is not systematic classification, but evolutionaryclassification. A top-down approach has shown to be far more reliable as it subsumes multiple and conflicting phylogenetic hypotheses.[53] But this is not how vernacular classifications work.[54]

Folk taxonomy is hardly rigorous, and since the words often preceded the science, a degree of revision based on science is inevitable. The technical name "dinosaur", however, entered the English language after they were named in 1843 by Richard Owen. So the claim that birds are dinosaurs is a case of a folk taxonomic term, bird, (one that agrees closely with scientific usage) being subsumed under a technical term, dinosaur. The claim that humans are apes is less simple. In folk taxonomy, "ape" is a term that has no comparable scientific meaning. It basically means any primate that lacks a tail and is not human, a double privation.[55] "Human", however, denotes a single and scientifically accepted species (or group of species), so here the claim is that the technical taxon falls within a folk taxonomic category. This is not new, of course, since Linnaeus famously placed humans (*Homo*) within the same genus as other apes, a classification that was later changed to reflect folk taxonomic preferences. Now the claim is that humans (*Homo sapiens*) are apes (Homininae), which is a group defined as the African Great Apes. In short, it is a claim that humans are a species of African Great Ape (and therefore also a member of Hominoidea, which includes gibbons and orangutans, sometimes also included among the apes).

Folk taxonomies are not always vague or socially determined. When the objects being classified are things that the language users need to be accurate about, because they hunt, farm or otherwise employ them, folk taxonomies can be quite exact and natural. Mayr famously recounted that out of 137 species of bird "European naturalists" (i.e., Mayr) identified, the tribe in the Foré Mountains of Papua New Guinea that he was visiting identified 136, and the one they disputed was also disputable in western taxonomy.[56] However, when Edward O. Wilson repeated the question about ants, he had less success.[57] The reason appears to be that locals must know birds well if they are to find and hunt them, whereas ants are rarely of economic significance. So folk taxonomies do not carry much weight when addressing natural classification, but neither should they be deprecated. They are often based on a lot of experience, most especially when, as with hunters and taxonomists alike, there are economic considerations in play.

Notes

1. Good 1935.
2. Scerri 2007, xix.
3. This is widely accepted now, following Lakatos 1970, 1978; and it is the core of what has come to be known as the Social Epistemology movement. However, there is a tendency in that tradition to speak as if all that mattered in science were power relations and political alliances; science is also about learning about the world independently of our political interests. Indeed, our political interests, and our competition, are among the engines that drive this process (Wilkins 2008).
4. See Lewis 1969. Conventions are unintended outcomes of attempts to coordinate behaviors.
5. Nelson 1978, 105.
6. Hull 1988, 81, citing Lindroth 1983, 1.
7. See his Appendix F, 527 ff.
8. E.g., Losos, Hillis, and Greene 2012.
9. HandMade Films, 1979.
10. Toulmin 1970.
11. Stevens 1994.
12. Which he called "descent with modification".
13. 1983, 1990, 1991.
14. This is an overgeneralization. Library science often discusses natural classifications under the rubric of "logical classification".
15. Magnus 2012.
16. Much of this was discussed extensively by 1948, but subsequent discussions dismissed it, as he noted, as "advocating a return to pre-evolutionary

thinking", just as previous discussions in German by Kälin and Naef had been dismissed.

17. And always has been. You can find people like Owen, Agassiz and T.H. Huxley asserting this.
18. In biology, a specimen is usually a whole organism, although it need not be. We are not presupposing that all biological specimens are whole organisms.
19. C.f., Barker 2010; Bird 2009; Colless 2006; Devitt 2008; Gayon 1996; Ghiselin 1974; Griffiths 1996; Hull 1965, 1978, 1988, 1992, 1999; Kitcher 1989; Kitts and Kitts 1979; Levit and Meister 2006; McOuat 2009; Nelson 1985; Okasha 2002; Ruse 1969, 1987, 1998; Sober 1980; Walsh 2006; Wilkerson 1993; Wilkins 2010; Wilson 1999a, 1999b; Winsor 2006a, 2006b.
20. Griffiths 1999.
21. Turner 2002.
22. Boyd 1999a, 1999b, 2010; Ereshefsky 2010b.
23. *Essay* II.xxv.7.
24. Whitehead and Russell 1910; Russell 1919. See also William Hamilton's discussion (Hamilton, Mansel, and Veitch 1874, v. 2).
25. Locke (*Essay* II.xxviii.2) held that the words that name relations, like "father" were conventional, but the actual relations, the genealogical relations, were not. See Wilkins 2009b, 65.
26. Mayer, Wallenius, and Ray 2005.
27. Redfearn 2002.
28. Hennig 1965, 21. Ironically, Hennig took this from the literature on "hermeneutics" in the social sciences.
29. Stephen Jay Gould has written about this often (Gould 1977, 1981, 1988, 1997, 2002). He notes that evolutionary trees are often progressionist, and occasionally racist. To what extent evolutionary trees must be these things is unclear, contra Ruse 1996. One argument, that there is an anthropocentric left-to-right bias (Sandvik 2009) is not convincing.
30. see Anderson 1976; Bonnet 1745; Pallas 1766.
31. Bentham 1827; Whately 1875. See McOuat 2003 for a discussion of this.
32. Stevens 1983. Lamarck's diagram is not a classification tree, but is better understood as a kind of "road map" that separate lineages can traverse as they are impelled by the *feu éthéré* to progress.
33. Herdman 1885.
34. On Hitchcock, see Archibald 2009; on Bronn, Gliboff 2007, Junker 1991. A general treatment of trees in classification is in Williams, Ebach, and Nelson 2008, chapter 4. On trees in language, see O'Hara 1996.
35. Hennig 1950, 208.
36. Watt 1947.
37. Hamilton 1967.
38. Whitehead 1938. This is not a fallacy of reasoning, formal or informal. It is a philosophical error rather than a logical one, like the naturalistic fallacy in ethics.
39. Marcuse 1964.
40. Sterelny and Griffiths 1999, 97ff.
41. Wilkins 2003, 2007a, 2007b.
42. Putnam 1981, 51.
43. Cracraft 2000.

44. The massive literature on names in philosophy is, we believe, largely irrel-
    evant to the nomenclature problem in the natural sciences, for whatever
    solution one adopts, either an epistemic referential account, or a historical
    referential ("baptismal") account, will turn out to be true of all names in
    science anyway. However, in practice, names in the historical sciences, and
    particularly the biological sciences, tend to be baptismal names, applied to
    a holotype ("type specimen"), and they thereafter remain unchanged no
    matter how the definite descriptions change. See LaPorte 2003; Levine 2001;
    and Rieppel 2003b for differing views on this.
45. Based on the synonyms in Riedman 1991. With the naming of new genera,
    this is widespread across all groups.
46. Chris Glen informed Wilkins of this. Anatomical nomenclature of the same
    group, such as trilobites, sometimes differs even between countries.
47. Hull 1988.
48. Leibniz 1996, 319.
49. Linnaeus 1758–1759.
50. Owen 1843.
51. The convention is that cladograms are rooted at the base (or at the left).
52. Sensu Hennig 1950; Mayr 1949, etc.
53. Williams, Ebach, and Nelson 2008.
54. Here the authors have a disagreement. Ebach thinks that this is a mistake
    and that it conflates folk taxonomies with systematics, and is to be avoided.
    Wilkins believes that language is pliable and influenced by science, and that
    the folk taxonomies have come to reflect the scientific ones. Hence, Ebach
    objects to calling birds dinosaurs and Wilkins is in favor of it.
55. Barbary "apes", which are macaque monkeys, do have stubby tails. The name
    was given before clarity appeared even in the folk taxonomy.
56. Mayr 1969, 313.
57. Wilson 1995.

4
Homology and Analogy

I'll teach you differences... [King Lear Act I, Scene IV]

Instead of producing something common to all that we call language, I am saying that these phenomena have no one thing in common which makes us use the same word for all, – but that they are related to one another in many different ways. And it is because of this relationship, or these relationships, that we call them all "language". [Wittgenstein, *Philosophical Investigations* §65[1]]

After having experienced the circulation of the blood in human creatures, we make no doubt that it takes place in Titius and Maevius. But from its circulation in frogs and fishes, it is only a presumption, though a strong one, from analogy, that it takes place in men and other animals. The analogical reasoning is much weaker, when we infer the circulation of the sap in vegetables from our experience that the blood circulates in animals; and those, who hastily followed that imperfect analogy, are found, by more accurate experiments, to have been mistaken. [Philo, in David Hume's *Dialogues in Natural Religion*, Part II]

Science arises from the discovery of Identity within Diversity. [W.S. Jevons[2]]

In this chapter, we consider what homologies and analogies are, in biology and other contexts. A homology is a relation from one set of objects or parts to another, a relation of identity no matter what differences of appearance or function exist in the parts or objects. Similarity relations are arbitrary, while homological relations are not. Homological relations are inductively projectible, based on consensuses of topographical agreement over time (causes) and space (forms). In biology, phylogenetic classifications are partial solutions to Goodman's grue paradox, since homologies are the right dependence relations to make inferences from. This is why, although the evolution of species is somewhat grue-some, we can make ampliative inferences about organisms. We finish with a discussion of Sober's "modus Darwin" being based on convergences (analogies) rather than the homologies (affinities) Darwin actually employed.

What is homology?

"Homology", as a term, arose first in mathematics, where it initially meant a mapping relation between sets of things, such as points in geometry. It was used by many people in various ways, but the founder of the modern use in biology is Richard Owen, who took the prior usage by Geoffroy Saint Hilaire of "analogy" and the discussions by Macleay and others of "affinity", and came up with the term in 1843,[3] and explicated it in detail in 1848.[4] He wrote

> The corresponding parts in different animals being thus made namesakes, are called technically 'homologues.'; The term is used by logicians as synonymous with 'homonyms,' and by geometricians as signifying 'the sides of similar figures which are opposite to equal and corresponding angles,' or to parts having the same proportions: it appears to have been first applied in anatomy by the philosophical cultivators of that science in Germany. [Owen 1843: 5]

The term itself basically just denotes the mapping of a relation from one set of objects to another; the relation is called homology and the things related are called homologs (or, in the Anglophilic spelling, homologues). We can think of this more concretely as like the relation between maps. A map at a scale of 1:50,000 maps onto a map at a scale of 1:10,000; the items represented on the smaller scale map are also represented on the larger, even though there are more elements in finer detail, and the elements are themselves of different specific forms.

Now consider the famous Belon diagram (Figure 4.1). Each labeled element (bone) in the human is mapped to a similarly labeled element in the bird. However, some of these bones are fused in the bird, and some are separated where the human bones are not. Homologies are mapping relations of shared elements – traits, or more properly, characters, since

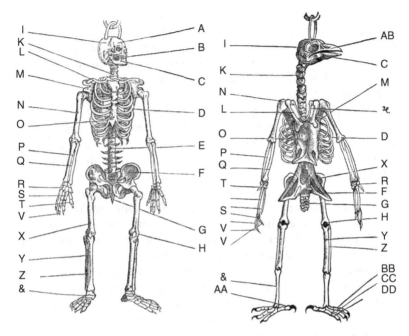

Figure 4.1   Pierre Belon's mapping of human and bird skeletons in 1555, Wikimedia Commons.[5]

the relations apply between abstract objects (characters are the names of traits; character states are the descriptions of characters). This is critical: the homologies are abstract relations. We shall consider what homologous relations are later, but at the moment it is enough to say they are formal, not concrete, mapping relations, between abstract descriptions.

The key aspect of homological relations is that the mapping tells us what are the same elements in each set. Such elements need not be "similar" in some respect, nor must they be in the same "place" in the set. Consider a rotated picture of a Spitfire airplane and a model of that plane. If the model is shown at a different angle to the real plane, we could still draw mapping relations between them no matter how different each element – say, the wing – looks in the two-dimensional image. If the model were of a later or earlier version, it may lack features on the real plane. It almost certainly lacks some of the features of the real plane (such as the rivets and small elements of the real plane too hard to make at that scale), and vice versa (real Spitfires lacked glue seals and extruded plastic bits inside). And so on. But the relevant elements are there in both, and if we describe each physical object at a suitable coarseness, we can say that they are in an objective homological relation to each other. Homology, then, is the relation between abstract objects (descriptions, or representations of real world objects) where the formal description allows a mapping function between them.

What, then, is analogy in this context? It is clear that classifying in any manner involves classifying by mapping from one set (a set of physical objects for our purposes) and so it must involve isomorphisms of some kind. Analogous relations are still a kind of homomorphism, but the mapping is not between sets of objects, but between the form of the individual objects themselves, and form is a pretty amorphous notion. The philosophical literature is rife with discussions on "similarity" and "resemblance" and usually starts off with the comment that in some way or another, everything resembles everything else.[6] So we need a formal notion of what "resemblance" or "similarity" consists in. Nelson Goodman declared similarity to be an ill-defined notion, and there is an extensive literature in the psychological and computer science fields as to what counts as similarity. Not surprisingly, there is also a large literature on it in taxonomy and computation. As a rough cut, there are three main approaches: Hamming Distance, Edge Number and Tversky Similarity. For now, let us understand classification by analogy as a similarity relation between things, sets, and forms.

Classification by analogy is sensitive to the metric chosen, but also to the representations of the things being analogized. For example, if we classify two organisms as predators, we are representing only a very small number of properties of the two organisms, and they are the properties contained in the definition of "predator" – one species that eats another. We do not represent almost any of the rest of the properties of the taxa being classified this way. One way to make this point, and show the difference between homology and analogy is to ask, what is it that the classification tells us?

If, in biology, someone tells you that X and Y are predators, all you know about them is already contained in the definition of "predator", and nothing else. There are exemplary predators, like lions and eagles, of course, but so too are single-celled organisms that engulf others, as well as fungi that predate on ants, plants that predate on insects, and so on. You do not even know if they are motile or sedentary, because there are "wait-and-catch" predators.

Contrast their telling you that two organisms are "Raptors". This is the taxonomic sense of "Raptor", not the popular sense of Jurassic Park; it refers to a particular group of birds known as "birds of prey" (in particular, the Falconiformes).[7] Now, if you are told that X is a Raptor, what do you know? An enormous amount. You know that it has a beak (a recurved beak!), claws (recurved also, and very strong), feathers (including flight and tail feathers of a particular structure) and that it has a particular diet (meat) and lifecycle (lays eggs, parents them, builds nests, mates in single pairs, is territorial) and so on. In short, what you know from studying several raptors is generalizable to all others, in a non-grueish kind of way. This also applies to as-yet-undiscovered properties. If we discover that Raptor X has enzyme E, then we can (defeasibly) infer that all other members of the Raptor group have E as well! That's an enormous amount of inductive warrant and return on cognitive investment. Interestingly, if we tell you that a raptor is a predator, you cannot infer that all raptors are (some, Old World vultures, are scavengers). Homology does not license analogical claims. But it may bracket them, as we will later argue.

We can summarize the difference here by saying that classifications by homology are inductively projectible, while classifications by analogy are deductive only.[8] Moreover, analogies are generally model-based. The choice of what properties to represent usually depend upon some set of "pertinent" properties, and this is not derived from an ignorance of what matters, or some unobtainable theory-neutrality. In order to measure similarity, you need to know what counts. The problem with the phenetics school of classification was that it failed to specify what counted, and so it got inconsistent results depending upon what principal component axes were used.

As we noted above, Locke showed there were infinitely many relations between things. Plato noted that there are at least some resemblances between any two things:

> Well, at any rate, he said, justice has some resemblance to holiness; for anything in the world has some sort of resemblance to any other thing. Thus there is a point in which white resembles black, and hard soft, and so with all the other things which are regarded as most opposed to each other; and the things which we spoke of before as having different faculties and not being of the same kind as each other – the parts of the face – these in some sense resemble one another and are of like sort. In this way therefore you could prove, if you chose, that even these things are all like one another. But it is not fair to describe things as like which have some point alike, however small, or as unlike that have some point unlike. [Plato, *Protagoras*, 331d–e]

Compare this text from a modern Plato, Nelson Goodman

> Similarity, I submit, is insidious. And if the association here with invidious comparison is itself invidious, so much the better. Similarity, ever ready to solve philosophical problems and overcome obstacles, is a pretender, an impostor, a quack. It has, indeed, its place and its uses, but is more often found where it does not belong, professing powers it does not possess. [Goodman 1972: 437]

We need to be very careful with likeness, similarity, resemblance and other (similar?) ideas when doing anything conceptually in science, because it is so very easy to find similarities. If you aren't careful, you will make inferences based on your own dispositions about the natural world; this is anthropomorphism, making the world in your own image, and is what science must overcome to be science. It relies upon the onto-logical fallacy previously discussed.

What, if anything, is similarity? More importantly, what is a similarity relation? Given how much of natural classification is claimed to depend upon it, we must ask these questions early on. In taxonomy, and we gather also in semantics, the similarity of one thing with another is roughly the Euclidean distance between them when they are mapped onto a representational state space. By this we mean (or think the specialists mean) that one takes all the variables in play and sets up a dimension for them each, forming a "feature set". Then one applies the particular value of each thing as represented (say, as measured) and makes the sum of these values the coordinate of that thing in the space constructed from the dimensions.

Obviously, these dimensions can be of a very high number, so for illus-tration let us suppose there are only three variables. The "location" of A in that space is the ordered triplet $<x, y, z>$ where each variable has a value. This is a Cartesian coordinate. Now if you have another object B and you want to know how similar (or inversely, how dissimilar) it is from A, you just measure the diagonal distance according to a formula some-times called the Hamming Distance, which is the modulo of the value between them. Sometimes many dimensions are collapsed into Principal Components and the distance is the summary distance between these, but that's a matter of tractability and computational convenience only.

Another way is to treat the axes as being discrete, and the similarity/dissimilarity is the minimum number of steps it takes to get from one to the other. These steps form graphs, and each coordinate is a node in that graph, for which reason we call it the node-edge definition; but it is a special case of the Hamming Distance version. It is also called the "nearest neighbor" metric or the "taxicab distance" or the "city block" distance, though such distinctions do not affect us here. This has the advantage of being easier to compute and represent (because you can draw the network graph in two dimensions), and may be more realistic about how people often estimate similarity relations given that we tend to gather things into discrete classes. However, it is at best a psychologistic convention, and tells us nothing much about the natural similarity of things. This might indicate the core problem with similarity itself, as Goodman pointed out.

Ironically, since we are trying to understand natural similarity and not psychologistic similarity, one of the clearest expositions of the similarity relation comes from Tversky and his collaborators, developed in the field of semantic and cognitive psychology. In trying to work out how people identified multivariate forms, for example, faces, as similar, or semantic notions like "fork" and "spoon", Tversky worked out a measure of similarity (and conversely also, dissimilarity), which is very abstractly described below.

Take a set of properties, a list of salient features (the "feature set"). The similarity relation is the intersection of some subset of these properties that two objects have. Each object to be compared has a set of properties taken from the feature set; call them set A and set B. Similarity is then a function of the mapping of the members of the sets A and B onto each other, according to this formalism:[9]

$$S(a,b) = \theta f(A \cap B) + \alpha f(A-B) - \beta f(B-A) \quad \theta, \alpha, \beta \geq 0$$

In short, to find the similarity between each set, one weights each set (the Greek letters represent the weighting), and the set of A minus the unshared properties of B, and the set of B minus the unshared properties of A, and subtract them from the weighted shared properties of both, and this gives you the similarity between the two objects A and B denote. A less formalistic version is:

$$\text{Similarity}_{AB} = \text{Shared}_{AB} - \text{Unique}_A - \text{Unique}_B$$

The dissimilarity is the inverse of this: the sum of the two unshared sets minus the shared set. This is also called the feature contrast model of similarity, and as such ties nicely into a contrastive account of explanation.[10]

Several interesting things follow from what is, we believe, the best general definition of similarity on the market.[11] One is that the degree of similarity is a matter of the choice of salient features or properties.

As we know, there are an infinitely large number of properties in common or potentially in common between any two objects. What we choose as salient will depend a lot upon us and our dispositions. Why, for example, is the square of the number of electrons in each object not used as a similarity metric? Because we do not have easy access to that information and it is unclear how interesting that sort of similarity would be anyway. Still, the number, and its exponents, are facts about the objects. That we do not choose them as salient tells us more about ourselves than the objects, and a Laplacean Demon may find that number or class of numbers very important indeed, in ways we cannot envisage.

The second point to note is that Tverskyan similarity tells us nothing about the objects that we did not put into the measure in the first place by setting up the feature set and the values applied. It helps us understand what humans are doing (or computers if they employ this metric and method; say, when searching text for semantic similarities), but it isn't extra information.

We have not explored phenetics hitherto, but now is the time to do so. This school of thought, which was very popular under the name "numerical taxonomy", arose with the rise of available computers in the 1960s and 1970s.[12] It aimed to deliver "theory-free" taxonomies by the mechanical application of (Hamming-like) algorithms to plain and atheoretical data. It seemed like classificatory objectivity was finally in our grasp. However, the methods, while mathematically rigorous and useful in many contexts, did not deliver the desired atheoretical taxa (which they called "operational taxonomic units" or OTUs, to avoid prejudging ranks like species).[13] Or rather, it delivered way too many; change the principal components and you got different taxa. Moreover, a Hamming Distance metric requires that you arbitrarily choose a threshold value to delimit the clusters. So bacteriologists, for example, tended to choose a 70 percent similarity or clustering value, while other biologists selected a 90 percent, 95 percent or even 99 percent value. This arbitrariness again resolved down to our predilections.

Nevertheless, while Hamming similarity was not a good way to identify taxa, it was a great methodology for identifying and analyzing clustering of various values, such as sequence or even genomic similarity in molecular genetics, and the algorithms are part of every taxonomists and bioinformaticians' toolkit today, quite rightly. What we need to understand is not what the metric is, but why it is useful and what it implies. "Phenetics" has become something of a dirty word in some circles, and that is a pity. It's like saying that because we cannot find exact definitions in language, we have to impose them (which is, we fear, what some philosophers do indeed say when confronted with vagueness).

Phenetic classification, and its analogues in other sciences, are classification by analogy; rather sophisticated analogies, to be sure, but analogous reasoning nonetheless. We select which analogies to employ (the feature sets), and so we have loaded our inferences from the beginning. When such inferences are called for, that is not problematic. When we think we have discovered something about the natural world we didn't already know, and all we have done is analyze our own dispositions, that is when the errors start to creep in. Similarity is deductive. It doesn't license inductive projectibility. It is not the foundation for inferences about history unless we can find causal inheritance – that is to say, identity relations or conservation relations. We will discuss below how one philosophical account, by Elliot Sober, has erred in just this way.

In short, homology is an identity relation, not a similarity relation.[14] It is a mapping of parts to parts in other entities.[15] Simply saying that a homology is a mapping function from one representation or model to another is not enough to give us a natural classification, however, for there are an indefinite (and very probably an infinite) number of ways to map two continuous models to each other. In biology, for example, we might map elements in many different ways, simply because biological organisms are composed of many very small elements that can be divided up, conceptually, for description in many ways (cells, which are composed of subcellular structures, which are composed of molecular objects such as proteins, and so forth). So we can reasonably ask: how do we arrive at homological statements?

In biology there have been many suggestions, including (wrongly) that homology is inferred from similarity, but this cannot be right, because homologous parts can be greatly dissimilar, morphologically, functionally and even developmentally.[16] However, there is a resemblance of one kind between homologs: structure. In the Belon diagram above, the "same" bones are identified because there are the same kinds of topological relations between, say, a maxilla in a bird and a maxilla in a human. The maxilla of a human is rounded, and holds teeth, while in the bird it is a toothless pointed beak, but it is located in the same relation to the bones of the skull, and the rest of the organism, and so it can be identified on that basis alone. But first, you need to be able to map the topological structure of the class in which birds and humans are located (which is, roughly, vertebrates). How can we identify that structure? Rainer Zangerl discussed this in 1948,[17] concluding that we do two things: one, set out the structural plan, which is the most general schematic of parts in that group even if not all members of it have them; and two, set out the morphotype, or more specialized schematic of parts for the subgroups. So in our Belon case, we identify all the parts vertebrates are made from (including parts not found in one or both of birds and humans) and their arrangements, and then we identify the particular parts and arrangements for birds, and the parts and arrangements for mammals, etc.

Once you have done this, we can start to identify the mapping relations between taxa within both the wider structural plan and the more restricted morphotype. We can simplify this diagrammatically:

By establishing the broader group's structural plan (sometimes referred to as the *Bauplan*, a German term that has the unfortunate implication of a builder's blueprint, with unpleasant idealist implications for some) and the more specific morphotype, inferences can be made on the basis of the overall affinities of subgroups. To return to our raptor example, if you find that hawks and eagles are distinct in some aspect, say, claw curvature, then some intermediate unobserved raptor cannot be said to either have one state or the other (although it is likely to have one or an intermediate form).

Homological identities are usually thought of as relations of morphology, or the structural form of the organisms under analysis. However, homological relations include mappings of sequence in development as well. Since Karl Ernst von Baer's observation that more

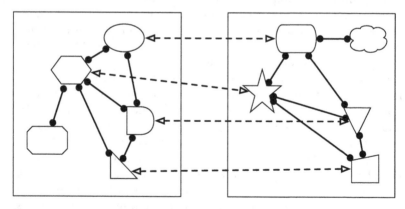

Figure 4.2 The topological relations between two objects allows us to map parts even though the parts do not closely resemble each other, and not all parts are mappable. The topology can be spatial, temporal, or even formal relations of parts so long as they are empirically forced. Hence a common sequence of the development of parts in organisms can form a homological relationship. This is sometimes also called the "topographical" notion of homology.[18]

general groups tend to share early developmental sequences while more restricted groups share later sequences (and individual species tend to have their own quite late developmental unique sequences), it has been the case that developmental sequence is itself to be understood as a kind of homology.[19] This suggests a way to generalize the notion of homology to sciences that have no simple analogue to development, for example, geological taxonomy. The topological relationship here is one of causal sequence,[20] and so long as the etiological process that generates a kind of mineral or soil, or a particular outcome in some other science, exhibits the "same" sequence, that is to say the sequential topology is commensurate, we may apply the notion of homology outside the developmental sciences. However, an account of what is "the same" is required.

If similarity is based upon the choice of feature contrast set, identity should not be. Identity relations are relations of exact identity, not general similarity. How does Belon's pigeon maxilla function as the same as the human maxilla? After all, if you replaced the pigeon's beak with a human jawbone, the result would not be pretty. The answer has to do with what it means to be taxonomically identical.

Broadly speaking, taxonomic identity is intersubstitutability. Leibniz' Principle of the Identity of Indiscernibles, which initially applied to substances rather than objects (that is, to classes of identical things), suggests that the epistemic test for some thing being identical to another is that it is not discernibly difference. But in biology and the special sciences generally, identity is never total, so discernibility relies on what information we have and what degree of resolution or fine-grain we are dividing the domain into. Emanuelle Serelli suggested in conversation that identity relations in a domain are in effect insensitivity to the properties to be used to classify. Choosing some different criteria to evaluate should give you the same answer – X is a member of the kind A whether you use criteria p or criteria q. However, similarity relations depend exactly on the criteria or properties used. Complete similarity relies on an absence of variability. Identity relations do not.

To flesh this out – one cow is "the same" as another in an ecological context because its trophic relations are insensitive to its kin relations and coloration, and individual quirks. It is "the same" in a reproductive sense because it can be exchanged with another cow without much difference in the reproductive outcome. However, in the context of phylogeny, the criteria used (which may be genetic or morphological) matter very much, and so identity (of species) will depend upon a consensus of traits. The indiscernibility here is one of taxonomic relations (as indicated by biological homologs).

Homologies in biological systematics come in all kinds of shapes and sizes. What they have in common is that they are relations between parts of objects (homologs) in taxa. We have to distinguish then between the homolog itself or themselves (the parts of a particular object in a taxon), and the relation between two or more homologs (which is the homology) between taxa. The literature on what counts as a homology in biology is vast and contentious.[21]

Soil Classification

An example of a natural classification outside the biological sciences is that of pedology, or the study of soils.[22] Pedology is primarily morphological and comparative, and seeks to view soil as a natural entity, with an evolutionary history. It has a century and a half of debates over classification parallel to biology.

Soils are made up of a source rock, sediment, organic matter and an original depositional environment, which together, determine the nature and classification of soils. For instance, in an undisturbed environment where there has been no ploughing or other human disturbance, it could be possible to determine the evolutionary history of the soils themselves. This idea led nineteenth century Russian soil scientists to propose a genetic classification of soils based on the types of environments where they found (i.e., subboreal humid forest and meadow soils).

However, most soils worldwide are disturbed, meaning that many of the original environments for are unknown (e.g., deforestation, back-burning etc.). Moreover, similar soils may have very different independent origins – for example, the Dark-Prairie soils of Alabama and Mississippi look the same, but are formed by two very different environments. A breakthrough came with the works of US pedologists G.N. Coffey and C.F. Marbut, who created a classification based on soil composition and morphology rather than on soil forming processes, creating perhaps the first natural classification in soil science. The appearance of new types of classifications has become a national past time. The US Soil Taxonomy is vastly different to the genetic-evolutionary Russian system, in the same way that Linnean taxonomy is different from Buffon's classification.

The similarities between pedology and biological classification are closer than one would think. Soils, for example, are treated as taxa, with different origins and history, but which can be compared by their morphology (i.e., soil color, texture etc.). Moreover, known environmental processes, that cause chemical leaching, erosion, and so on can also be taken into consideration. In this way, soil science becomes systematic and comparative as well as evolutionary-genetic, which focuses on the soil forming processes of specific types of soil.

Given that each of these classifications are endemic to the country in which they are used, most classification schemes have never had to compete in the same way as biological classifications. The soil types of the United States, for instance, cease abruptly at the 49th parallel to be replaced by the soil types, with partial correspondence, of the Canadian system.[23] Moreover, each country tends to have unique soil types, meaning that the Russian system would not work well in Australia for instance (e.g., Australia has no arctic tundra). Perhaps the national adoption of different classification systems for unique soil types has prevented soil science from developing universal classification syntheses like those in the biological sciences.

Intriguingly, pedology has named genera and species of soils, but not in a fashion that closely mirrors Linnean nomenclature.[24]

Induction, and the straight rule of homology

For inductive inferences to be successful, we have to guard against the grue problem.[25] While this is very familiar to philosophers, it is less well-known to biologists, so a short summary is in order. The grue problem is based on a kind of "broken" predicate or property: ordinarily we might infer from the fact that all prior emeralds have been observed to be green and the future emeralds would also be green – this would be a case of inductive inference. But in the absence of prior certainty about what the rules are, without a "straight rule",[26] we cannot rule out the existence of another property, which Goodman calls "grue", in which emeralds are green if observed before some time *t* and blue thereafter. Hence, every observation of a green emerald strengthens the inference that after temeralds will be seen to be blue. It must be noted that this is not a claim that emeralds will change color. It is about what we can infer of unseen members of a class. A gruelike predicate is unprojectible, which is to say that it cannot be projected to unobserved entities. What inductive inference promises is projectibility, so that we can say things that are very likely to be true about unobserved members of the class.

This is more than a mere thought experiment. The infamous "black swan" example so beloved of logicians is a simple case. Swans were all observed to be white, until black swans (*Cygnus atratus*) were discovered in Australia in the late seventeenth century. Had swans been defined by a white plumage (which was common at the time[27]), then swan plumage would have been a grue property. More generally, consider such taxonomic classes as Mammalia. Mammals are defined as animals which lactate, and have hair. But whales have secondarily lost their hind legs and (most) hair, not everything that apparently "lactates" is a mammal, and monotremes lay eggs. More recently it has become clear that lineages of species are not straight but "grue-some". What defines a group of organisms may change in a daughter species. Consider sexuality as a trait of a group of lizards such as the genus

Aspidoscelis (whiptail lizards). When one species becomes partheno-genic (secondarily asexual) we encounter a grue property for real. This applies to any single property of a group, potentially. Evolution leads to grue problems.

And yet, biology is not deeply troubled by grue problems, even though it is precisely the science that should be. While the color of the swan's plumage turned out not to be projectible, the new black swan was not placed in a new order or class. It was recognized to be a swan nevertheless, and placed into the existing monotypic genus.[28] Although philosophers, who anyway tended then as now to rely upon folk taxo-nomic categories for their examples, were shocked in the manner of Captain Renault, biologists simply shrugged, reported the new species, and added it to the existing taxonomy. The reason is quite obvious by now: the swan was not defined, but classified upon the overall affinities it exhibited, and the fact that one homolog differed in character state from the rest was not crucial, any more than if it had a different shaped beak from the rest of the genus.

The issue here is with what Godfrey-Smith calls the "dependence relations":

> We should not make a projection from a sample if there seem to be the wrong kind of dependence relations between properties of the sampled objects.[29]

It is our claim here that homological affinity does act to provide the "right" kind of dependence relations between properties of taxa. A single failure of a homolog to project properties is insufficient to make the taxon unnatural (that is, in philosophical terms, an unprojectible class), since the class (the taxon) is formed from the overall suite of homological relations (which we are calling the affinity, following early nineteenth century taxonomic use). Affinity acts to set up taxonomic kinds, and these act as a "straight rule", as they do tend to converge upon projectible properties.

In taking inference from homology to be a kind of "straight rule", the question is why it works. If the universe were such that properties correlated by chance, it would not work, but in the cases of the special/palaetiological sciences, properties correlate due to a shared productive cause. If the universe lacked appreciable structure of this kind, then no search method would deliver knowledge (cf. Wolpert and Macready's "No Free Lunch" theorem[30]). Assuming that properties can be corre-lated, the epistemic question is how to identify those that are and to distinguish them from those that aren't, which in biological systematics is the distinction between homology and homoplasy. If there's knowledge to be had, then one way to acquire it is to iteratively refine one's classifications in an attempt to maximize the homological relations on which they are based.

Such inferences are, of course, quite defeasible. It should not be thought we are supposing that natural classifications are in any way certain, or that any given homolog will exemplify the same states in each taxon or object classified. Of course this will not apply. On the one hand this is probabilistic[31] inference, in the sense that there is some likelihood or confidence that the projection will succeed for each property, and a high confidence that it will succeed for most properties. This is akin to selection on a smooth landscape versus on a rugged landscape; selection can act on traits in a highly correlated "smooth" landscape (where adjacent coordinates are not too different in value from each other), but it fails on an uncorrelated, or "rugged" one.[32] The progress of science has been compared to an adaptive walk,[33] and similar considerations apply to inference in science as apply to selective searching of the adaptive landscape; both are special cases of a search procedure of the kind Wolpert and Macready discuss.

Systematics in biology, and classification in science generally, resolves much of the practical issues of grue-some induction by, as Godfrey-Smith says, ensuring that the right class is sampled by finding the right dependence relations through a process of iterative refinement. These are what we are generally calling homologies. We will discuss the particular kinds of homologies in different sciences, such as serial and transformational homologies in biology, in later chapters.

Modus Darwin and the real modus darvinii

Elliot Sober has published a claim that Darwin used, and we should too, a particular syllogism: similarity, ergo common ancestry.[34] This cannot be right for three reasons: logical, historical and inferential. First the logical, as this is rather vapid, and can be guarded against (although Sober does not[35]) relatively simply: it cannot be that similarity in itself is evidence of common ancestry, or every dice would have a common ancestor, and every rock that resembles Abraham Lincoln's profile would share a common ancestor with old Abe. The way to guard this might be to assert that yes, they do have common "ancestors", in the general sense they have common etiologies. All dice resemble each other because there is arguably a chain of cultural descent that links back to some "dice taxon" in the past, maybe some-where in Asia. The rocks have a shared etiology in the recognition of the physiognomy of Abraham Lincoln. But that is not quite the claim Sober is proposing. For this would involve the cognitive and cultural dispositions of ourselves as classifiers, and common ancestry in no way relies upon us, although of course our recognition of it does. Can we infer from similarity that the two objects that are similar (to us) have a shared causal history? The Lincoln case suggests not. One rock might be formed by a volcanic extrusion, while another might be half a world away and formed by the deposition and lithification of sediments. Without limitations on the kind of similarity, it implies nothing at all about the objects (and perhaps quite a lot about the observers engaging in pareidola).

The historical objection is that Sober, and most other modern commen-tators, read Darwin wrongly. Darwin used not similarity, but affinity, as evidence for common ancestry, and technically, he inferred common ancestry from "group subordinate to group" taxonomy; that is to say, he explained this taxonomic arrangement with common ancestry, rather than defended the claim of common ancestry that way. Had he wanted to use similarity, there was a perfectly good term, before Owen's inven-tion of the notion of homology: analogy, as can be found in the discus-sions in the Quinarian literature. Darwin wrote, in chapter XIII of the first edition of the *Origin*:

> all organic beings are found to resemble each other in descending degrees, so that they can be classed in groups under groups. This clas-sification is evidently not arbitrary like the grouping of the stars in constellations. [411]

And he goes on to note

> Naturalists try to arrange the species, genera, and families in each class, on what is called the Natural System. But what is meant by this system? Some authors look at it merely as a scheme for arranging together those living objects which are most alike, and for separating those which are most unlike; or as an artificial means for enunciating, as briefly as possible, general propositions, – that is, by one sentence to give the characters common, for instance, to all mammals, by another those common to all carnivora, by another those common to the dog-genus, and then by adding a single sentence, a full description is given of each kind of dog. The

ingenuity and utility of this system are indisputable. But many naturalists think that something more is meant by the Natural System; they believe that it reveals the plan of the Creator; but unless it be specified whether order in time or space, or what else is meant by the plan of the Creator, it seems to me that nothing is thus added to our knowledge. Such expressions as that famous one of Linnæus, and which we often meet with in a more or less concealed form, that the characters do not make the genus, but that the genus gives the characters, seem to imply that **something more is included in our classification, than mere resemblance.** I believe that something more is included; and that propinquity of descent, – the only known cause of the similarity of organic beings, – is the bond, hidden as it is by various degrees of modification, which is partially revealed to us by our classifications. [413f, emphasis added]

Darwin goes on to discuss how external resemblances are not evidence for propinquity (nearness, or kinship). He discusses how similarity is mere "adaptive or analogical characters" and that it is "a general rule, that the less any part of the organization is concerned with special habits, the more important it becomes for classification". Darwin knew well about convergence. "We must not, therefore, in classifying, trust to resemblances in parts of the organization", he concludes. That we need an ensemble of characters, and that they are not necessarily about similarity, is clear from this passage:

The importance, for classification, of trifling characters, mainly depends on their being correlated with several other characters of more or less importance. The value indeed of an aggregate of characters is very evident in natural history. Hence, as has often been remarked, a species may depart from its allies in several characters, both of high physiological importance and of almost universal prevalence, and yet leave us in no doubt where it should be ranked. Hence, also, it has been found, that a classification founded on any single character, however important that may be, has always failed; for no part of the organisation is universally constant. The importance of an aggregate of characters, even when none are important, alone explains, I think, that saying of Linnæus, that the characters do not give the genus, but the genus gives the characters; for this saying seems founded on an appreciation of many trifling points of resemblance, too slight to be defined. [417]

And he then discusses affinities by saying "Our classifications are often plainly influenced by chains of affinities" [419]. Affinities, not analogies (and as we argued, "affinity" means roughly shared sets of homologs). He summarizes by noting that

> All the foregoing rules and aids and difficulties in classification are explained, if I do not greatly deceive myself, on the view that the natural system is founded on descent with modification; that the characters which naturalists consider as showing true affinity between any two or more species, are those which have been inherited from a common parent, and, in so far, all true classification is genealogical; that community of descent is the hidden bond which naturalists have been unconsciously seeking, and not some unknown plan of creation, or the enunciation of general propositions, and **the mere putting together and separating objects more or less alike.** [420, emphasis added]

It is plain that Darwin held that what was evidence for common ancestry was shared sets of homological relations independently of adaptive characters, which can converge. Affinities, not analogies, are evidence and Darwin knew this well.

This brings us to the inferential objection. Sober fails to deal with convergent evolution as a cause of similarity, and yet this is so well-known to systematists as to be hardly worth discussing. Because he adopts what is basically a statistical notion of classification, Sober thinks, we suppose, that homoplasy, which is usually interpreted to mean convergence[36], is eliminated somehow by technique or methodological algorithms. However, every systematist strives to minimize non-homology before analyzing data, just as Darwin said. There is no magic method for doing this: what looks homological may turn out, upon comparison of many taxa, to be homoplasious or indeterminate, and vice versa. But despite our limitations here, we can do this successfully in most cases – if we could not, then we could not do natural classification at all.

In neither place where Sober advances modus Darwin, does he defend against this obvious objection. In conflating similarity with affinity, we are confused about what counts as evidence for a given scenario of common ancestry. Although we have suggested that there is no fixed or privileged direction of inference in a field, it does appear that if you begin with uncertainty, then recognition of naive classification based on homological relations is going to constrain and set up the explanandum for the hypothetical account to explain. The hypothesis, a historical narrative, is not evidence for itself.

Darwin is often used as a heroic figure upon whom the preferred philosophies of the writer may be painted. In that respect he is like the heroes of the Bible, except that he is a lot clearer as to his intent. The actual inferential process Darwin used – the real *modus darvinii*[37] – is more like this: affinity, explained by common ancestry. Since affinities are groups of homological relations we might use a modern term of Hennig's and say that synapomorphies give the pattern that the historical process, in Darwin's view, explains. The two are not identical.

If anyone is responsible for the form of inference that Sober ascribes to Darwin, it is Diderot.[38] He wrote:

It seems that nature has taken pleasure in varying the same mechanism in a thousand different ways. She never abandons any class of her creations before she has multiplied the individuals of it in as many different forms as possible. When one looks out upon the animal kingdom and notes how, among the quadrupeds, all have functions and parts – especially the internal parts – entirely similar to those of another quadruped, would not any one readily believe (*ne croirait-on pas volontiers*) that there was never but one original animal, prototype of all animals, of which Nature has merely lengthened or shortened, transformed, multiplied or obliterated, certain organs? Imagine the fingers of the hand united and the substance of the nails so abundant that, spreading out and swelling, it envelops the whole and in place of the human hand you have the foot of a horse. When one sees how the successive metamorphoses of the envelope of the prototype – whatever it may have been – proceed by insensible degrees through one kingdom of Nature after another, and people the confines of the two kingdoms (if it is permissible to speak of confines where there is no real division) – and people, I say, the confines of the two kingdoms with beings of an uncertain and ambiguous character, stripped in large part of the forms, qualities and functions of the one and invested with the forms, qualities and functions of the other – who then would not feel himself impelled to the belief that there has been but a single first being, prototype of all beings? But whether this philosophic conjecture be admitted as true with Doctor Baumann

[Maupertuis[39]], or rejected as false with M. de Buffon, it can not be denied that we must needs embrace it (*on ne niera pas qu il faille l'embrasser*) as a hypothesis essential to the progress of experimental science, to that of a rational philosophy, to the discovery and to the explanation of the phenomena of organic life.[40]

Note that Diderot is making an argument for classification based on transformations of forms (which he called prototypes), whether or not that is explained by actual historical processes. Nevertheless, the inference here is clearly *similarity ergo ancestry*, even if Diderot is merely posing it as a way of saving the phenomena in the manner of a Ptolemaic astronomer. Darwin did not use that inferential principle, as by that stage, homology was recognized as an identity relation, not as a measure of similarity.

Conclusion

Homology and analogy, being derived from biology, in which there are lineages of descent, may not be all that useful for a general account of classification, and yet there are underlying commonalities in all classifications in the special sciences. They all rely upon contingent events and sequences, and they all have time-relative and boundary condition-relative causal processes. It is of course a truism that each scientific domain will have its own special factors in play when classifications are done. A mineralogical classification, a soil classification, a psychological classification and a biological classification are all going to be their own thing. And each of these will differ in epistemic kind from inferences made in a general science, of if one is a true physicalist, in physics, which is the only general science (relegating chemistry to an epistemic domain of physics).

But classification is more than just a local epistemological activity. It sets up the problem sets for a new domain, to be sure, but it also primes the ontology of that domain, even if later that ontology is reduced or explained in terms of another theoretical structure or resource. We shall discuss what makes a domain of investigation its own field rather than just a way-station on the way to a real reductionist destination later. For now, it is worth asking what benefit lies in the distinction within a domain between homology and analogy.

Fundamentally it is a distinction between deduction from a prior model or explanation, which suggests we do know something about the domain in question, and inference to as yet unknown taxa. We will never be so ignorant of a domain that we know nothing to begin with (or else how do we even know it is a domain worth investigating?), but we must allow that we can do some discovery, or else what is the point of doing empirical science? We might as well just run mathematical simulations all the time if discovery is prohibited by a failure of inductive inference. Hence, the etiological aspects of classification being the foundation for inductive inferences is rather crucial. When we find ourselves ignorant of a domain, and just to the extent we are ignorant, classification by homology is the way out of the fly bottle. When we know something, and just to the extent we do know something, about the domain, deductive inference from our models and explanations is one way to classify. Since we are never in a situation of complete ignorance nor complete knowledge, we will always need to constrain homological classifications by some prior knowledge, and to constrain analogical inferences by homology. Classification in this sense is the foundation for progress in science.

Notes

1. Wittgenstein 1968, 31e.
2. Jevons 1958, 1. Original edition in 1873.
3. Owen 1843, 379.
4. See chapters 7 to 10 in Williams, Ebach, and Nelson 2008 for a historical overview of the use of "homology"; also Jardine 1967.
5. Belon 1555, 40–41.
6. A point made by Locke, *Essay*, I.XXV.vii, among others. It's probably in the scholastics; cf. Plato below.
7. Wilkins particularly likes this group because of the role they played in species concepts, via Frederick II's book *The Art of Hunting with Birds*, as outlined in Wilkins 2009b, 39–42.

   However, John Harshman has unkindly and unjustly destroyed this philosophically neat example by pointing out in correspondence that the group "Raptors" has now been divided into two disparate clades, the order Falconiformes, which include around 290 species of diurnal birds of prey, and the order Accipitriformes, which include around 225 species, including eagles, hawks and New World vultures (see Hackett et al. 2008). Though this makes the example less pleasing, it in fact strengthens our argument, as inductive inferences will now no longer be as projectible between them (just as the group "vultures" ceased to be so projectible when it was discovered that Old World vultures were more closely related to the Accipitridae, eagles and kites, etc., than to New World vultures).
8. We might say that in a philosophical context, homology gives us the reference of class terms and names, while analogy gives us the reference, but

while analogical classes have a sense, that is to say, an intension, not all denotations of class terms must be homological.

9. Tversky and Gati 1978.

10. Lipton 1990, 1991.

11. Wilkins is grateful to Michael Weisberg for the simpler formulation of Tversky similarity.

12. Sneath and Sokal 1973; Sokal and Sneath 1963.

13. The reliance of the pheneticist movement upon operationalism, a philosophical view that is most often associate with Percy W. Bridgman but which can be found in Cuvier, Newton and many others, is something not sufficiently investigated. Operationalists hold that the truth of a statement is irrelevant in science so long as it may be made "operational". Hence the rejection of, say, the biological species concept as being something that makes no operational difference (by, e.g., Ehrlich 1961) is a criticism founded upon operationalism.

14. There are special senses of terms like "homology" and "analogy" in other fields, and care must be taken not to confuse senses across disciplines or the abstract senses we are using with these terms here. For example in chemistry, a homolog is a repeated part of a molecule or a similar element, and this is considered part of an analog.

15. A view espoused by Ereshefsky 2007; Ghiselin 1997; Love 2007; Müller 2003, 205ff; Wagner 2007, who present the case that homology is identity. As Ghiselin says, "Analogy, like homology, is a relation of correspondence between parts of wholes. But in an analogy, unlike an homology, those wholes are not parts of some larger individual; instead they are members of a class" [207]. See also Ghiselin 2005, 97.

    Homology is a relationship of correspondence between parts (individual homologs) of individual organisms, which are in turn parts of individual genealogical wholes.

    A contrary view is given by Brigandt 2002, Brigandt and Assis 2009, who support the homeostatic property cluster view of Boyd 1999a, 2010.

16. Of course we identify the elements in terms of some similarity, based on our prior knowledge of the parts of the group being classified.

17. Zangerl 1948.

18. Cf. Franz 2005b; Nixon and Carpenter 2011. Geoffroy Saint Hilaire's principe des connexions is perhaps the earliest version of this made explicitly, Rieppel 1994. An early discussion of this is Jardine 1967, who also discusses a topographical model and resemblance.

19. Collazo 2000; Jaramillo and Kramer 2007; Laubichler 2000; Muller and Wagner 1996; Wagner and Stadler 2003.

20. Cf. Rieppel 1994; Wagner and Stadler 2003.

21. See Assis 2013; Assis and Brigandt 2009; Brigandt 2002, 2003a, 2007; Brigandt and Assis 2009; Brigandt and Griffiths 2007; Brower and de Pinna 2012; Cracraft 1967, 2005; de Beer 1971; Donoghue 1992; Fitch 2000; Freudenstein 2005; Ghiselin 1969, 2005; Griffiths 2006, 2007; Hall 1994, 1999, 2012; Hillis 1994; Hoßfeld and Olsson 2005; Inglis 1970; Inglis 1966; Jardine 1967; Key 1967; Laubichler 2000; McKitrick 1994; Mindell and Meyer 2001; Müller

2003; Nixon and Carpenter 2011; Opitz 2004; Patterson 1982a; Patterson 1988a; Pearson 2010; Platnick 2012; Ramsey and Peterson 2012; Rendall and Di Fiore 2007; Rieppel 1994, 2005, 2007; Scotland 2000; Shubin, Tabin, and Carroll 2009; Wagner 1989; Wagner 2007; Williams 2004; Williams and Ebach 2012. The evolutionary debate was begun by Lankester 1870 (see Gould 2002, 1071–1089).

22. Buol 2002; Buol et al. 2003.
23. Soil Classification Working Group 1998. The US scheme is more a natural system (Soil Survey Staff 1975) than the Canadian, which treats classification as a conventional system of natural properties.
24. Basinski 1959; Buol 2002; Marbut 1922; McDonald et al. 1984; Mückenhausen 1965; Soil Survey Staff 1975; Varfolomeev 2010.
25. Goodman 1954. See Godfrey-Smith 2003a for a discussion of the classical problem.
26. Hans Reichenbach proposed a "straight rule" for induction in his *The Theory of Probability* 1949, in which induction was justified when increasing observations converged upon an asymptote. See also Salmon 1991. Here we are using it in a more general sense, as a way of ensuring that gruelike properties are eliminated.
27. However, public houses named "The Black Swan" were in existence well before the discovery of actual black swans, possibly as an ironic title.
28. The northern swan genus is divided into four species, three of which are North American and were not named until the early nineteenth century. The black swan *Cygnus atratus* was named in the late eighteenth century. The Whooper Swan *Cygnus cygnus* was named by Linnaeus.
29. Godfrey-Smith 2003a, 579.
30. The "No Free Lunch" Theorem states that no single algorithm outperforms chance when amortized over all possible search spaces or functions (Wolpert and Macready 1995; Wolpert and Macready 1997).
31. Here we mean something like a likelihood probability. This may be Bayesian or some other statistical confidence; it doesn't materially affect the argument which philosophical stance towards probability one adopts here, and we leave it to the reader to convert the argument to their favorite method or position on the matter.
32. Gavrilets 2004; Kauffman 1993.
33. Hull 1988; Wilkins 2008.
34. Sober 1999, 2008, §4.1, 265ff.
35. Sober offers a number of statistical and Bayesian techniques, but so far as we can tell, none of them involve identifying homoplasy.
36. Homoplasy can be interpreted to mean convergence. But it could also mean bad data. In molecular systematics it may refer to xenology (see Chapter 8).
37. Wilkins is indebted, yet again, to Reed Cartwright for help with the Latin here.
38. Nelson 2011. We are greatly indebted to Gareth Nelson for this and much other information. He was doctoral advisor to both authors, and we benefitted greatly from his tutelage.

39. Pierre Maupertuis published *Venus Physique* pseudonymously in 1747, in which the first scientific theory of evolution was published.
40. Denis Diderot, 1753, *Pensées sur l'interpretation de la nature*, ch. XII, as translated by 1904, 325. Gary Nelson pointed this passage out to us in a preprint of his cited article above.

5
Monsters and Misclassifications

DELTA: I cannot understand how an able man like Alpha can waste his talent on mere heckling. He seems engrossed in the production of monstrosities. But monstrosities never foster growth, either in the world of nature or in the world of thought. Evolution always follows an harmonious and orderly pattern. [Imre Lakatos[1]]

Logic sometimes breeds monsters. [Henri Poincaré[2]]

... either Fiction, or want of Observation has made more Monsters than Nature ever produced. [John Floyer, 1699[3]]

He who fights with monsters should look to it that he himself does not become a monster. [Friedrich Nietzsche[4]]

In this chapter we consider monstrous classifications, or misclassifications, which rely more upon facts about the observers and their predilections than upon the facts about the objects classified. Trashcan categories are common in science, but are aphyletic in biological terms. We consider what is a natural classification, concluding that it is one based on a single cut of a classificatory hierarchy (monophyly in biology) rather than a mixture of artificial and natural characters. Natural kind classifications are grades based on the analogous characters preferred by a Theory.

Misclassifications

There are many ways to misclassify natural taxa, in addition to giving a classification by analogy and treating it as homological.[5] We need to have a general account of what is happening here. The most obvious is, of course, to simply classify arbitrarily for some non-epistemic purpose, such as arranging books on shelves by color. Convention classifications are just a matter of convenience, by definition. This includes our ability to retrieve records regarding the objects classified. Information retrieval is, in itself, a matter of data handling, and the same theoretical concerns arise whether you have data on 3 × 5 inch cards or when it exists in large data matrices on computers. Likewise, naming is not, in itself, a matter of natural classification, as we discussed earlier.

In his "Parts of Animals", Aristotle discussed one form of misclassification – privative taxa.[6] He wrote

> Again, privative terms inevitably form one branch of dichotomous division, as we see in the proposed dichotomies. But privative terms in their character of privatives admit of no subdivision. For there can be no specific forms of a negation, of Featherless for instance or of Footless, as there are of Feathered and of Footed. Yet a generic differentia must be subdivisible; for otherwise what is there that makes it generic rather than specific? ... From this it follows that a privative term, being insusceptible of differentiation, cannot be a generic differentia; for, if it were, there would be a common undifferentiated element in two different groups.

His argument is based on a logical (predicate or term-based) notion of essentialism that does not concern us here, but the general point – that you can get many contrary classifications from privative groups – survives even his subsequent argument that

> The method then that we must adopt is to attempt to recognize the natural groups, following the indications afforded by the instincts of mankind, which led them for instance to form the class of Birds and the class of Fishes, each of which groups combines a multitude of differentiae, and is not defined by a single one as in dichotomy.[7]

Aristotle is proposing that Fishes and Birds are natural groups. Modern biologists might agree for birds (Aves, in the Linnaean system, Avialae in one modern retelling) that they are a single group, but not fishes, since all terrestrial tetrapods would need to be included. Ironically, Aristotle's classification, and indeed the entire folk taxonomy that he refers to as "the instincts of mankind", is a classification by analogy, in that he takes functional properties like "living in water" as the differentiae, or characters, to construct the class. In phylogenetic, or cladistic, taxonomy, this is not sufficient. As we discussed in the last chapter, functional properties can be evolved through convergence (homoplasies). And yet, taxonomists and philosophers continue to attempt to classify by metrics of functional, morphological and analogical similarity, insisting that what cladistic terminology dubs paraphyletic taxa are real groups, informative, and necessary.

In this chapter we shall consider these sorts of classifications, both inside biology and without. They are a kind of "monstrosity". The need to form monsters, and our subsequent need to prevent them from interfering with our scientific inferences, is crucial to understanding natural classification.

Monster making and monster barring

In his classic dialogue, *Proofs and Refutations*, Imre Lakatos considered what the right classification of geometrical figures and polyhedra were. He has his interlocutors attempt several definitional, or in our context essentialist, schemes, only to find that there are exceptions to all these. One student, Delta, complains of the Socratic approach used here

> But don't you see the futility of these so-called refutations? "Hitherto, when a new polyhedron was invented, it was for some practical end; today they are expressly invented to put at fault the reasoning of our fathers, and one will never get anything more from them than that. Our subject is turned into a teratological museum where decent ordinary polyhedra may be happy if they can retain a very small corner."

A footnote refers to Poincaré's similar remarks concerning what a function is. The "Teacher" remarks

> I think we should refuse to accept Delta's strategy for dealing with global counterexamples, although we should congratulate him on his skilful use of it. We could aptly label his method the method of monster-barring. Using this method, one can eliminate any counterexample to the original conjecture by a sometimes deft but always ad hoc redefinition ... [8]

Monster making and monster barring are perpetual activities in classification, whether or not the objects being classified are abstract or concrete. But natural classification always classifies concrete objects, and for all but possibly fundamental physicochemical entities, no definitions are feasible. The targets of Lakatos' deliberations include Popper and Feyerabend, but they have a broader import.

What makes a monster? In general terms a monster is something that fails to fit into what should be an exclusive, comprehensive or possibly a "natural" classification scheme. However, every classification scheme has exceptional cases, called "trashcan groups" by systematists. Linnaeus himself initially had to call a group of unclassifiable monstrous taxa in his first cut at the Empire of Nature in the *Systema Naturae* in 1735 as "Paradoxa" (Figure 5.1). No matter what definienda or differentiae are used, in any sophisticated and open system, some things just won't go (and on that basis Linneaus dismissed all but one, Satyrus, as myths, and that one as probably an ape if travelers weren't mistaken)[9]. The goal of a natural classification is to ensure that every item being classified has a positive not a negative (that is, a privative) set of differentiae. To achieve this, the method used needs to ensure that trashcans are not filled, or if filled, not named and treated as real.

Monsters tell us very little. A monster is, almost by definition, something or some group that is not natural. However, often what is counted as "natural" is something like "congenial to the intuitions of the folk", as Aristotle said, or even to the intuitions of the researcher. Basically a monster is something that transgresses intuitive categories. In the Middle Ages, obsessions about monsters had a strong moral dimension, and it seems this persists even now.[10] Taxonomists sometimes insist upon the reality of monstrosities and their informativeness, when the reality is that what monster taxa inform us of is our own dispositions. A monster taxon is a surprise, something that contradicts our prior expectations.

Monsters in the sense of the Lakatosian discussion are undefinable exceptions. If our classification scheme is based upon definitions (or essences, in the proper sense of being definable predicates, not in the substantive sense of there being some causal essence, as analogical schemes are), then there must be exceptions in any plausible domain of investigation, because unless we have a complete theory for the domain and for any object, process or dynamic that might influence or enter into the domain, there are going to be items that do not fit the scheme. In contrast, a homological scheme might be exhaustive, and have no residual items, since what makes an item something to classify at all is the possession of homologies, and as these are positive properties, all groups constructed on that basis will be positive rather than privative groups. To understand this, let us consider the role of "primitive" groups in biological systematics.

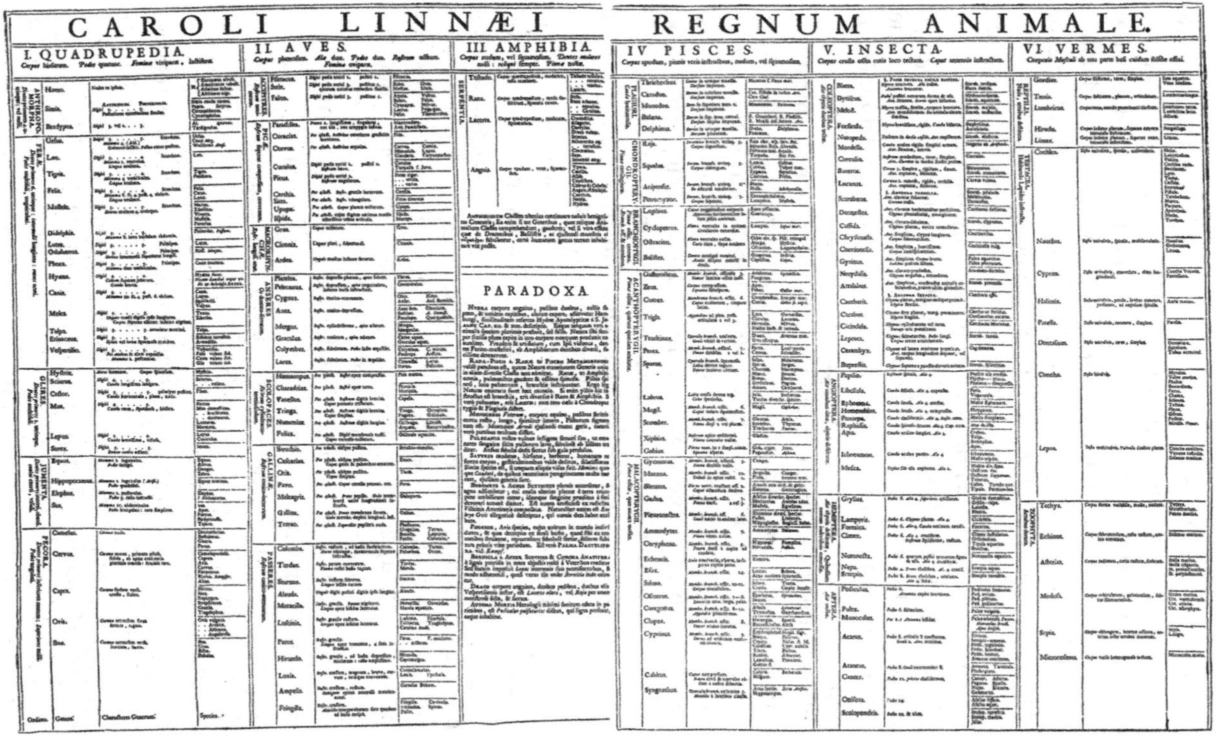

Figure 5.1 Linnaeus' initial table of animals. The Paradoxa are those species or groups that share characters of several taxonomic types, although some are merely mythical – phoenix, dragon and sartyr, the rhinoceros/unicorn, the hydra (which was problematic right up until the end of the nineteenth century) and a frog-fish. Wikimedia image

Progress, primitive and plesiomorphy

One of the more difficult conceptual problems the layperson has with biology lies in the use of the simple word "primitive". It has many antonyms – "modern", "evolved" and "derived" – and like many biological uses of ordinary words, everybody thinks they understand it, and don't. It is a word from the Latin, of course, for "first fruits" or "first things of their kind", but in modern use it means "simple" or "undeveloped". And this is not – quite – what it means in biology. It got its modern sense via the early nineteenth century social philosophy of August Comte, who thought that societies evolved through predetermined stages – the theological, the metaphysical, and the scientific (thus starting the movement known as positivism). His ideas impressed a legal professor at Cambridge, Henry Maine, who concluded that society had passed through definite stages, from "primitive" to "progressive"; and these ideas spread from there.[11] Progress was then, as now, in the air. Things could only get better, and society was forced to improve. No sudden reversals, only improvement. It is obvious, therefore, how such ideas, which collectively go by the name The Great Chain of Being,[12] would be transferred over to the living world when evolution, or transmutation as Darwin called it, got applied to life. What was true of the political world had to be true of the living world also. This is exactly how the two founders of evolution over time had applied the Great Chain – Jean Baptiste de Lamarck, and Erasmus Darwin, Charles' grandfather. For Lamarck and Erasmus Darwin, life began simple and unorganized, and got more complex and sophisticated in what it could do. But Charles Darwin and Alfred Russel Wallace had a different notion of progress. In fact, they had only a local notion of it. Things, if they got better, did so locally and relatively. And they could do badly too: they could go extinct, a possibility not acceptable under the Great Chain (the good God would never permit part of his creation to disappear). Groups of organisms, or taxa, as they came to be known later, could become hard to find. There could be, in other words, a dearth of taxa.

It is famous that Darwin used a tree diagram (Figure 5.2) to represent evolutionary relationships. But tree diagrams can mislead. Darwin knew this – he had himself documented barnacles that became simpler as they evolved, not more complex, and he wrote to himself in a margin of a book "never say higher or lower" (self-advice he disregarded, by the way). We popularly tend to think more like Lamarck or the first Darwin and think that something earlier in the evolutionary tree was somehow "not as good" as the modern taxa in that tree. But all that is required in

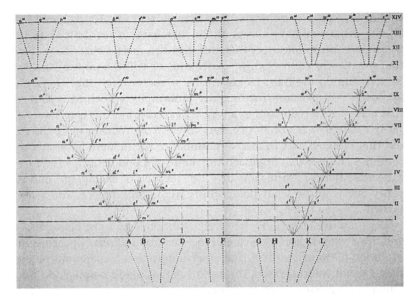

Figure 5.2 Darwin's diagram in the Origin Darwin 1859, 116–117

Darwin's scheme is that each new species is better adapted to local conditions than the older ones. There is no insistence upon transitivity – if each species is better adapted at local conditions than its predecessors, it doesn't follow that the last in a series of species is better overall than the first in the series. Each species is well-adapted to its own conditions.

In the mid twentieth century, the German entomologist Willi Hennig proposed a way to classify taxa that was deliberately founded on Darwin's scheme of evolution.[13] He used the characters of organisms and arranged them so that the more inclusive forms and traits could be seen as those of the ancestors. Hennig coined many words from Greek to replace older and confusingly ambiguous terms. Two in particular were important. Hennig defined the ancestral character with an adjective: plesiomorphous. This Greek term means "neighboring form", because it is shared by all the neighbors on the tree. Any character that is derived from the plesiomorphy is a version of it, but is changed. These he called apomorphous, from the Greek that means "away from" + "form". Apomorphies defined branches on the evolutionary tree that had, by Ockham's Razor, likely changed later than the plesiomorphies that all the rest of that tree shared. Hennig's terms remove the implications of Comte's view of history or the Great

Chain. Evolution is now seen as a change from one form shared by all the other branches except those that have forms derived from it. There is no implication that this gets the species that share that trait a little closer to God, or perfection, or white European males. By focusing on the traits of organisms, some have felt that Hennig's approach is a reversion to the older philosophy known as typology that preceded Darwin, and in a way, that is true. But this is not necessarily a flaw. The historian of taxonomy M. Polly Winsor has pointed out that typology was in fact very similar to the modern approach of finding examples that approach the mode of a population, which we call the mesotype.[14] She calls this the "method of exemplars" and it really does capture what taxonomists were doing before, and after, Darwin.

So, "primitive" has been effectively abandoned by most biologists,[15] and those who do use it, are careful to avoid implications of progress in evolution.[16] However, there is another implication to plesiomorphies: they are not informative.[17] The reason is that plesiomorphies do not differentiate groups, whereas apomorphies do.

Paraphyly and plesiomorphies in biology

When Hennig formalized phylogenetic classifications, one of his key influences was John Henry Woodger, a logical positivist who attempted, many years ahead of his time, to axiomatize biological classification.[18] Woodger's theory involved representing taxa as sets (the formal appendix of his book was prepared by Alfred Tarski, which is no small thing), in which the enclosing taxon set represents the last common ancestor of the taxon sets it encloses. Given the nature of proper sets, this meant that for something to be a taxon, it must be a proper subset or superset of other taxon sets. The complement, or remainder of the superset once the proper subsets have been taken out, is not a taxon. It is that trashcan again, and the hope is that once we have properly represented all the taxa in a superset, there will be no remainder. That, at least, is the ideal. Whether we can do this in practice, or whether we have sufficient information to do it in principle, is another matter.

Hennig called the complement of a superset, once the proper subsets were removed, *paraphyletic*, a term that he coined. There is another form of improper setting he called *polyphyly*, roughly the inclusion of partially intersecting sets in a single superset. Paraphyly and polyphyly are the inclusion in a group of only some of the shared upstream nodes of a

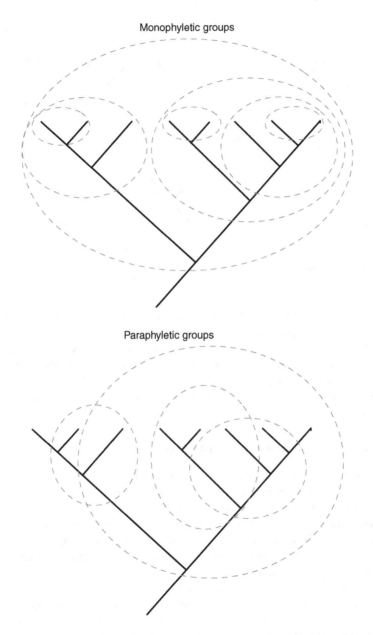

Figure 5.3 Hennigian natural and unnatural groups: Monophyly, and the aphyletic groups paraphyly and polyphyly. All monophyletic groups have a single "cut" of the cladogram. All aphyletic groups have two or more

Polyphyletic groups

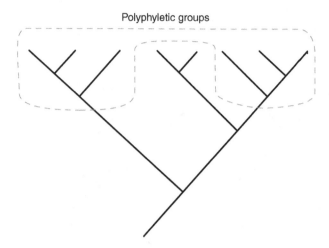

Figure 5.3   Continued

deeper node, and the inclusion of nodes with unshared deeper nodes, respectively (Figure 5.3). They are regarded now, as a result of Hennig's influence and theory, as "unnatural groups", or more exactly, as "artificial groups". Paraphyletic and polyphyletic groups, which we may together call *aphyletic* groups, are, in effect, monstrosities of classification, and rely upon facts about observers rather than facts about the taxa.

Hybrid classifications and phylogenetic bracketing

Systematists often generate hybrid classifications, where classification by clustering (clouds), by similarity (grades) and by shared etiology (clades) are mixed together and inconclusive inferences made upon that basis. In this instance, the kinds of inferences that are licensed by homological classifications are not licensed commensurably by analogical classification, and so one commits a fallacy of warrantability in drawing the one from the other (Figure 5.4).

There are three kinds of classificatory activities, in general:

Artificial classifications are merely conventional. They may be adopted for operational, communicative, or merely traditional reasons. In biological systematics, it is agreed that artificial taxonomies are of limited value and do not contribute to theory development except incidentally.

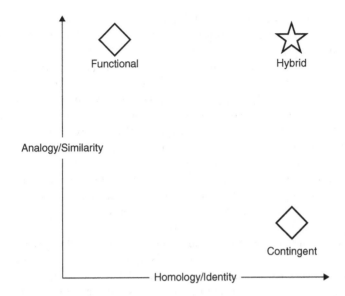

Figure 5.4 Hybrid classifications mixing homological and analogical classification schemes

Natural classifications properly and independently of human dispositions divide the world into its structural components. This is the asymptotic ideal for taxonomies in science. However, since there is always some human element in the taxonomy, perforce, it need not be realizable in order to be important to science.

Hybrid classifications arbitrarily mix artificial and natural classifications, which can, if care is not always taken, lead to invalid inferences and deteriorating research programs. Hybrid classifications are common in biology, although it is not generally agreed upon which are hybrid, and which are not.

Nevertheless, hybrid classifications are rife in biology and other special sciences such as psychology. If they are to be used, how can they be used properly? Mixing analogies and homologies is always fraught with peril. One way to do this is to do what has been called phylogenetic bracketing.[19] Initially this was a technique to make inferences about the lifestyles and soft tissues of extinct fauna by bracketing them with their nearest living relatives and drawing (inductive) inferences from the known taxa. Here, however, we mean it to be something like this: in order to make analogy-basedinferences of taxa in a special science, we should first constrain the domain of the inference to those taxa which are very closely related. Since we are talking about extending this beyond biology, let us call it etiological bracketing. A very nice example of this approach was made by Christopher Glen, who measured the claw curvature of many bird species in order to cross compare lifestyles of birds (arboreal, terrestrial or mixed) with those of dinosaurs with which they are closely related.[20] The reason the inference has weight is because birds and theropod dinosaurs are so closely related they likely shared the same developmental and behavioral properties. Comparing this to, say, pterosaurs, would not work so well, and to primates not well at all.

Since history and causation tends to eliminate analogical relations, etiological bracketing must restrict inferences in the history-based palaetiological sciences to closely related taxa (Figure 5.5).

By constraining analogous inferences we can avoid, for example, the profligacy and unwarranted inferences made by sociobiologists in the 1970s, where the behaviors of deer ("stamping grounds"), bees (individual sacrifice for the superorganism), and birds ("pecking order") were uncritically applied to human behavioral dispositions. Just to the extent these taxa were related to humans did the inferences have any warrant, which is to say little at all, leading to unjustified claims about the adaptive worth of this or that trait. By contrast, consider the inferences made from primates to hominids by Frans de Waal.[21] Here, since we share "primitive" traits, which is to say ancestral properties not otherwise modified by evolution, some inferential warrant applies well, pace future discoveries about uniquely human evolutionary novelties (so few and far between).

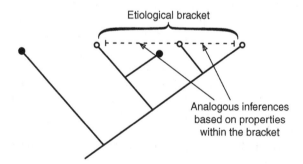

Figure 5.5   Bracketing analogous inferences within homological classes. Notice that the ampliative inference within the etiological bracket is defeasible, and indeed in this schematic example is defeated in one case

Phylogenetic bracketing improves the worth of the inferences done within a group. Induction can curtail the excesses of similarity-based reasoning.

What is natural?

In this book we discuss natural classification, based on the model of modern biological taxonomy. In that discipline and methodological debate the notion of what counts as a "natural" group looms large, as it also does in such disciplines as soil science (pedology) and many other fields. Naturalness is a disputed property.

In phylogenetics, a group is natural if it is monophyletic. That is to say, if the cladogram is cut on a single vertex, all the upstream branches form a natural group.[22] Others who follow a gradist approach to a greater or lesser degree demur. Ashlock,[23] for example, wished to preserve the term monophyly for a different kind of naturalness based jointly on genealogy combined with grade attainment, and instead preferred the term *holophyly* for the cladistic monophyly, which he thought was artificial. What is at issue here?

There are many competing senses of "natural", and three are directly relevant in this case. The first sense is archaic, and was once preferred by Francophone biologists: "agreeable to the mind". A natural concept was one the mind naturally preferred. In this sense teleological conceptions are "natural". We can put this sense to one side, so long as we note that people often mean by natural that something is easy to conceive or use, even when they are overtly appealing to one of the other two senses. When observers note phenomenal patterns, and on that assert their reality, they are actually asserting facts about themselves, that these patterns are agreeable to them; that is to say, they are salient to them as observer systems.

The second sense is "theoretical". Often, a kind term is held to be natural because it is a theoretically significant category. Functional terms, for example, are natural in this sense. Whenever a biologist or social scientist calls something "functional", they mean by this that it is a privileged theoretical property. To the extent that the ruling theory of a domain is accurate, true or explanatory, the kind is natural. The natural kinds literature generally relies upon this notion of naturalness, with the exception of Hacking's account.[24] Convergent grades in evolutionary biology, which rely upon the recognition of properties that have significance for theoretical reasons (like cognitive capacities, or homeothermy, or locomotory types), are theoretically natural. They are certainly objectively true features of the world, but they are privileged out of a multiplicity of features that might have been privileged, and the only justification for giving them such focus is the theoretical explanation to which the observer is committed. Theoretical naturalness is thus at least partly conceptually relative, based upon a choice of explanatory satisfactoriness.[25]

The third sense, and the one that cladistic monophyly appeals implicitly to, is "mind-independence". Something is natural if its being what it is depends in no way upon a mind or observer. To the extent that a classification represents mind-independent facts well, the single cut notion of monophyly represents structure that really is in the world. Where the cut is placed is arbitrary, but any single cut on a cladogram or tree diagram is like the proper set of set theory: it includes all the structure that is downstream of that cut.

Consider instead the aphyletic groups of evolutionary systematics: paraphyly and polyphyly. A paraphyletic group is a section of a cladogram or tree diagram that has two or more cuts, at least one downstream of the lowest cut. Since the choice of location for a cut is arbitrary, the remaining branches of the tree are arbitrarily grouped. We do not have all the structure represented in the diagram from the data used to construct it. So the choice of data is arbitrary and says more about the classifier than about the things being classified. A polyphyletic group is where cuts are arbitrarily made in different branches and the resulting group put together for reasons other than the structure in the data.

In both cases, the group formed is not formed on the basis of the structure in the data set, but upon some other choice of criterion, and this can only be a theoretical criterion. In short, grades formed from aphyletic groups are theoretically natural and hence at least partially subjective.

When these senses of naturalness are conflated, the result is a hybrid classification that does not license inferences made from them. A purely monophyletic classification licenses all kinds of inferences to unobserved specimens, traits or properties, but if theoretical considerations are introduced into the classification such ampliative inferences become deductive inferences. To this we now turn.

Floyer's monster pigs and his use of homologies to eliminate the interspecies breeding claim

In 1699, John Floyer reported upon a claim of interspecies breeding: human–pig.[26] He had shown to him by a breeder in Staffordshire a new-born pig with what appeared to be a human face:

Contrary to the claim made by some that this was due to interbreeding between humans and pigs, a view quite common in the literature on natural history from the Greeks through to the modern era, Floyer observes that the teeth, pelt, nose, and bones are generally those of a pig, and that the similarity to human faces is due to distortion "whilst the Bones were Cartilaginous" in

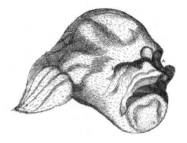

utero. He thinks it is due to mechanical distortion, but whatever we might now think of fetal deviation in development after such things as the thalidomide scandal, Floyer is using homology to overcome a mistake in classification made on the basis of analogy. He wrote

> I was further convinced in Opinion that there was really no mixture of the two Species in this Monster, by the Woman's account who saw the Sow take the Bore[sic], and after the sixteen Weeks, on the beginning of the seventeenth, which is the usual time, the Sow pigged [bore] eight Pigs, the first five were perfect Pigs, the sixth was the Monster, and after that two more perfect Pigs, all which I saw sucking the Sow, and as well shaped, and as large as possible...

Floyer is appealing not only to the species homologies of anatomy, but also to the species homology of development. He makes this explicit, when thinking of the Mule:

> the Female contains in her Eggs the first Rudiments of the Animal of her Species, and...the impregnation only changes some of the extremities into resemblance of the Male.

This is a view widely held at the time: the male changes but does not cause the inheritance of the species-typical traits. He goes on to argue that this is a general law of biological inheritance.

Assigning a monster to a single species is a remarkably modern thing to do for the late seventeenth century. Most monsters were thought to be hybrids from Aristotle's writing onwards. That he uses observation rather than theoretical presumptions is telling also, but clearly what is observed are traits that are held to be somehow informative as to the taxonomic relations between pigs and humans.

Notes

1. Lakatos 1976, 21.
2. Poincaré 2003, 125.

3. Floyer and Tyson 1699, 434.
4. *Beyond Good and Evil* §146, Nietzsche 1966, 89.
5. Note that classification by analogy, or theory-based and functional classifications, are not, in themselves, misclassifications of functional and theoretical objects.
6. Ogle translation, 642b22–643a5 (Aristotle 1995, vol. 2, 1000). See the illuminating discussion in Nelson and Platnick 1981, 70ff.
7. 643b10–14.
8. Lakatos 1976, 22–23. Thanks to Neil Thomason for directing Wilkins to this discussion, and forcing him to read it carefully.
9. Thanks to B. Ricardo Brown for the translation and commentary of the Paradoxa <http://until-darwin.blogspot.com.au/2012/11/the-paradoxa-from-systema-natura-by.html>
10. Daston and Park 1998; Haraway 1992; Park and Daston 1981.
11. Comte 1853; Mackintosh 1899.
12. Kuntz and Kuntz 1988; Lovejoy 1936.
13. Hennig 1950, 1966.
14. Winsor 2003.
15. Ayala 1988; Gould 1996; Mittelstrass, McLaughlin, and Burgen 1997.
16. But see Ruse 1996.
17. For example, John may be derived from his mother Mary. But when John has a daughter Sue, he instantly becomes a father or "primitive". While Mary, John and Sue are all derived, at some stage they become primitive. While derived is a constant condition of organisms, primitive is relative condition that depends on another derived condition and therefore is not necessarily informative about the group or characters.
18. Gregg 1954; Hennig 1950, 1966; Woodger 1937. See also Sandri 1969.
19. Witmer 1995.
20. Thanks to Chris Glen for introducing Wilkins to the notion of phylogenetic bracketing. Cf. Glen and Bennett 2007.
21. de Waal 1982, 2001; de Waal et al. 2006.
22. Baum 2008.
23. Ashlock 1971, 1972, 1979.
24. Hacking 1991, 2007.
25. When a classification is called subjective by systematists, it often relies upon an unforced choice of conceptually salient characters and properties relative to the subject matter. So "subjective" is a synonym here for "conceptually relative".
26. Floyer and Tyson 1699.

6
Observation, Theory and Domains

As all sciences are based upon facts, known, or to be known from experience, so are they, in their early state of developement [*sic*], matters of pure observation. It is only when we have acquired the power of generalising these facts, when such generalisations agree among themselves and with every thing we see or know of nature, that the theory of a science becomes either absolutely demonstrative, or approaches so near to certainty, by the force of analogical reasoning, that it is not contradicted by anything known. The case of natural history, then, is precisely this; in its early stages it is a science of observation; in its latter, it is one of demonstration. [William Swainson, 1834[1]]

In this chapter we consider how abandoning the full Theory-Dependence of Observation Thesis (TDOT) affects our view of classification. We define a scientific Theory (capital-T) as something distinct from the notion that phenomena are observed based on prior criteria of salience to an observer, and adopt the Bogen–Woodward notion of a phenomenon as a pattern in data. A phenomenon, including a classification, is the explicandum that Theory explains. We then consider the question whether Theory from outside a domain of investigation counts as theory-dependence within the domain, and thus ask what a domain is in science. We set up a "domain conundrum" – how can a science get started when there is no Theory of its domain?

Species, phenomena and data

In the first chapter we argued that classification is not something that depends upon theory. This implies a more empiricist account of classification than has been usual, so in order to explain how it is that classification can play an independent role in the dance of science, we need to consider one of the oldest chestnuts in the philosophy of science: observation and phenomena.

To do this, let us consider a question in the philosophy of biology: what are species? There are as many as 25 or so conceptions (not concepts – there is one concept being defined or described in different ways) in the post-1940 literature of species, and a number of philosophical accounts.[2] Not often considered is the role that species plays in biological taxonomy: is it a unit of classification only (or even at all?), or is it the unit of evolution, biodiversity or ecology? Most of the conceptions in the literature are technical definitions based on theoretical considerations, usually of population genetics. We wish to suggest that species is not a term of theory at all: species are phenomena that call for explanation.[3] We will use the question of whether species are phenomena as a surrogate for the question regarding all taxa in all sciences.

Several things are implied by this thesis:

Species turn out on this view not to be causal entities, but rather the epiphenomena of causal processes at the individual and populational levels. Species don't cause anything (apart from thoughts in the minds of taxonomists).

If species are phenomenal objects, what is a phenomenal object? Is it an ordinary object that

- is refined into a theoretical entity under something like entity realism?
- is divided into things that are theoretical entities once we get the right theories, leading to theoretical pluralism (or explanatory pluralism, depending how you parse that)?
- is deflated into a non-entity that is replaced with theoretical entities?

And just what is a phenomenon anyway? How is it different from data? (A distinction of Bogen and Woodward's.[4]) Is a phenomenon just an aggregate of data or is data one end of a continuum of which phenomena are the other end? Can phenomena be atheoretical?

The problem with phenomena is that they have been deprecated in modern analytic philosophy of science. They are supposed to be theory-dependent, based upon the language of the domain under investigation, and this idea goes back, as Michela Massimi has argued, to Kant.[5] Massimi asks whether we can presume that there are "ready-made" phenomena that have to be "saved" (Hacking has a gloss on the phrase "saving the phenomena"; the Latin is *salva*, which he makes cognate to the English "solve", thus to save the phenomena is to give an explanation for it[6]). This, she thinks, is the widespread or commonplace view – that science explains phenomena observed without theory, by theory. Kant and those who follow him (including the whole modern philosophy of science since Mach and Duhem), hold that we construct phenomena from data, a view reiterated by Bogen and Woodward.[7] Add to this the constructivist view that observation is theory-dependent – that we cannot even get data without prior theory – and we have a conundrum: in order to explain phenomena by theory we need theories which call forth the phenomena. Hence we do not ever actually discover phenomena in need of theoretical explanation; we make them. There are no "ready-made" phenomena.

If this doesn't seem counterintuitive, then you haven't read much history of science. The phenomena of planets was first observed without any theory of celestial spheres; indeed the latter were hypothesized to explain this (non-theoretical) phenomenon. Likewise, the various reactions of naturally occurring chemicals was observed before we had either the Aristotelian or the alchemical theories of elements, or before we had Daltonian theory, all of which were attempts to explain these naturally occurring phenomena. Granted, in physics and modern chemistry a good many of the phenomena are things that are made in experimental, controlled, settings, as Hacking insists (we intervene in the naturally occurring order to isolate causal influences). But there doesn't seem to have been a space left for unintervened phenomena, unconstructed by prior theory, in the philosophy of science. Why is that?

In part it is because of Kantian views, that we have access only to constructed phenomena and that ready-made phenomena might be indicative of confidence that we could see the *Dingen-an-sich* directly. This naive realism is not at all accepted by philosophers of science. A claim for the existence of natural phenomena unmediated by theory suggests that sort of realism, and so it is rejected. But does it imply naive realism? We do not think so. At best it implies that there are phenomena in the world we do not construct but can see (hear, smell, etc.) without there being theory that we base our perceptions upon. However, it is a long step from that to the claim that the phenomena are representations of the way the world is directly. Observing a chemical reaction (soda and lemon juice, for instance) is not something one can only do based on a theory of chemistry. It is enough that we have a commonsense experience of the world such that such a reaction is marked out as interesting and different.

A response often encountered is that in even observing these differences we rely upon theory: the "theory" that is embedded in our perceptual dispositions. Sometimes this is called psychologism, but in this case it is usually referred to as "evolutionary epistemology". These are the dispositions that we have to explore a "quality space" as Quine called it. The Kantian analytic a priori is the phylogenetic synthetic a posteriori; to modify a phrase of Konrad Lorenz'.[8] This is a markedly deflated notion of "theory", so poor as to mock the meaning of "theory" in science. If just being able to detect motion in a visual field is a theory, everything is theory and we need not attend to theory-dependence as such. But now we must still deal with scientific theory differently, and the phenomena that it explains. Let us call scientific theory, Theory, with the capital, and leave the rest to psychological and sociological dispositions. Is observation theory-dependent? Are phenomena Theoretical? We still say they need not be, although if they are constructed through intervention they may be.

Consider a case in which a scientist has an instrument that assays some property, like temperature. On the Bogen–Woodward view, the phenomena of the temperature of a heated piece of metal is the pattern of several measurements of that piece using the thermocouple (or whatever the tool is). Unless you have a pattern in the data, there is no phenomenon. Since the functioning of the thermocouple is explained by Theory (of, say, relative metal expansion rates), the observation is Theoretical. But any ordinary metal worker knows that metal heats up when put in flame, and they will have a pretty good idea of what you need to do in order to get a metal to just below the melting point, by the whiteness of the glow they give off (under uncontrolled lighting, too!); this is how swords were made for centuries. That is not Theoretical. It is just a ready-made phenomenon.

So there remains room for ready-made phenomena, even if to perceive them one needs to be trained and experienced, without Theory. Hacking makes something like this point about microscopes and telescopes – we ascertained that they worked without a theory of optics (even though some pretty good Theories of optics had been around since the Islamic Golden Age), by checking through direct observation that they were properly imaging objects, and making ampliative inferences from there.

To return to species: they began to be properly named at the *fin de siecle* of the herbalist medieval tradition, based largely upon experienced observation by people like Konrad Gesner, Caspar Bauhin, Konrad von Megenberg and others.[9] There was no Theory to speak of then that required species, apart from a need to track naturally occurring phenomena, and yet many of the species they named remain good species today. Moreover, some assay-driven splitting of species, based largely upon molecular techniques, is counterintuitive when the groups are split (into "Operational Taxonomic Units") using theory-driven criteria but run counter to broader observation of interbreeding, ecological adaptation, and so on. Theory-based instrumentalism is not triumphant by any means. We have in mind here the DNA barcodists who rely on a short stretch of mitochondrial DNA in eukaryotes, CO1, and similar sequences in "prokaryotes".[10]

Species can be identified by experienced observers in the absence of Theory. One must be trained and corrected in one's taxonomic apprenticeship, and much of the "knowledge" employed is tacit practice, gained, we suggest, from past experience of phenomenal distributions of closely related organisms.[11] The sense of Theory employed is ancillary, and not crucial (if one assay starts to fail, well, as Groucho once said, we have others). The theory-dependence of modern natural history (excluding, that is to say, physics and general chemistry) is not central to identifying the phenomena that stand in need of Theory to explain them.

The Theory of the assays, however, can range from distal and irrelevant (as the quantum theory of optics is for identifying, say, mammal species) to closely relevant and even partially overlapping the domain of observation (quantum optics may play a critical role in identifying microbial species with light microscopes, for example). We suggest that we should think of theory-dependence in a more nuanced way: a phenomenon is theory-dependent if there is no phenomenon observable unless a Theory is necessarily employed in the observation.

So particle tracks in physics are theory-dependent. However, most species observations are not (if we restrict ourselves here to species that can be observed in the absence of specialist tools like DNA barcoding chips). The characters and properties of species of fishes, flowers, and foxes are ready-made phenomena that can be seen if one learns how to, even in the absence of Theory, or else we cannot account for the history of taxonomy and natural history in general. The distinction made by van Fraassen between observing and observing-that is not even important in this context.[12] Moreover, such ready-made phenomena are a good reality check on the assay-driven phenomena that do rely to greater degrees upon Theory. Maybe one cannot see DNA clustering, but you can see morphological   traits (especially homologies) if you study the groups. If DNA clustering implies there is a phenomenon of species that we do not pick up by trained direct observation, that becomes a testable hypothesis; it is not true by assay definition. It might turn out that cox-1 is a highly variable gene in some species, or that it is shared across many species unchanged; such is the nature of evolution.

We therefore propose that we call "phenomena" patterns of observation, and index them to the theoretical domain in which they properly occur (that is, in which they are to be explained). If the observation does not rely on the domain's explanatory Theory, they are to be regarded as theory-independent for that domain. There will, of course, be Theory that explains why we can observe those phenomena (and it might be a distal application of the relevant domain: we see because we evolved to see objects of that kind, and evolution explains why there are objects of that kind – this is not to say that the observation is theory-dependent). But it is not necessarily a Theory of that domain. Many objects are observed because they are indeed objects of that domain. We are not denying this; we are merely saying that we have unjustly excluded from our philosophies of science a phenomenon of observing phenomena without Theory. One learns that by observing history (without a Theory of history, necessarily).

The periodic table

It might be thought that classification in the special and historical sciences is occasionally atheoretical, but that in the general sciences, physics and chemistry, it is derived from Theory. But in fact one of the most exemplary cases of empirical classification that led to Theory is in these sciences: the periodic table.

According to a study by Eric Scerri,[13] the standard textbook story of the periodic table is wrong, unsurprisingly since most such textbook narratives are so. The idea of a table of elements did not derive from Dalton's revival of atomism, but from increasingly refined laboratory techniques for ascertaining the atomic weight of elements. The notion of atoms, of course, did derive from Dalton but these developments did not seemingly rely much upon it. Because samples of element include what we now understand are isotopes, atomic weights are not perfectly correlated with atomic numbers. Consequently while it was apparent from the experimental data as the atomic weights of elements were refined that there was some sort of pattern, for example, by Stanislao Cannizarro and Alfred Naquet who both arranged these weights in a table, early attempts tended instead to rely on a kind of Platonist numerology, Johann Döbereiner's "theory of triads" in which patterns of elemental properties fell into threefold relations. These were not unlike the affinities of systematics, and indeed the term "affinity" was used by chemists. However, there were too many exceptions for triads to be the basis for a table of elements (although something like it later returned briefly).

Until Dmitri Mendeleev's table was published in 1869, the best previous version was that of Julius Lothar Meyer,[14] who, like Mendeleev had arranged elements in a series based on their atomic weights. What Mendeleev did better than Lothar Meyer and other precursors was to employ his vaster knowledge of the empirical properties of chemical elements, and to make an arrangement based on weights and properties, in a kind of family resemblance.[15] He refined it over many years, again based on the experimental results. The term "experimental" here is a little bit misleading, in our terms, because while there was intervention to refine and purify the samples used, this was not intervention to experimentally modify the samples along some independent variable to test a hypothesis. In our terms, it was simple empirical research.[16] Mendeleev even expressly noted that he was taking a Lockean or even operationalist approach:

> by investigating and describing what is visible and open to direct observation by the organs of the senses, we may hope to arrive, first at hypotheses, and afterwards at theories, of what has now to be taken as the basis of our investigations.[17]

Subsequent to the adoption of the table by chemists, there arose a program to improve and explain the "periodic law". As Scerri says (in the epigraph to Chapter 3), once scientists have a classification, they seek an underlying cause of the regularities (as Darwin did).

Following on from the adoption of atomic theory and the discovery of electrons by Thompson in 1897, and the nuclear structure of the atom proposed by Rutherford, Bohr's introduction of the quantum to the structure of the atom led to an explanation of valency, which was originally discovered by Edward Frankland and Friedrich Kerkule in the 1860s. Eventually, Rutherford and Antonius van den Broek proposed around 1911 that each atom had an integer number that gave it its weight. About the same time, the discovery of isotopes by Frederick Soddy explained why atomic weights varied: different samples had varying admixtures of the isotopes, whose weights varied.

In short, the periodic table is a mixed case of empirical research guiding theory development. It is not a simple example, because there was prior Theory, and also because much of atomic theory derived from other sources and cases, especially in the study of radioactivity, but nevertheless classification plays a key role in this most central of physical scientific domains.

The Diagnostic and Statistical Manual of Mental Disorders

This text, known by its acronym the DSM (-I, -II-, III or -IV, and -V published in 2013), is the main standard classification of mental illness and disorders used across the world, although it is primarily the production of the American Psychiatric Association (APA). It is not the sole classification, there being also International Statistical Classification of Diseases and Related Health Problems (ICD), produced by the World Health Organization. Both systems are widely used in the context of drug prescription, medical insurance and administration, government statistics, and medical education.

The original system that the DSM-I codified was based upon psychoanalytic categories.[18] DSM-I evolved out of the pre-war "Statistical Manual for the Use of Institutions for the Insane" after the Second World War, in 1952, relying strongly upon military terminology and practice during the war, especially the Navy's. Over the years it was revised extensively, and by DSM-III the APA abandoned the goal of earlier editors and authors to find an etiology for the diseases classified. Since very little in the way of etiologies for the diseases had been uncovered, it was seen that the role of the DSM was to provide practitioners with a way to efficiently and effectively diagnose conditions, and prescribe drug treatments and other treatments.

This meant that the DSM was not a nosology the way classifications of diseases were in medicine. Although medical science might not know what the etiologies of diseases were, the aim and project of medical research and classification was to move from phenomenology to etiology, and when a disease was finally explained in a way that might break it up into several distinct or more general conditions, medical science had little trouble doing so. Psychiatry, on the other hand seemed to move in the opposite direction. Instead of explaining and revising categories based on a knowledge of causal substrate, psychiatry revised based on "general concepts" of mental illness, some of which were in fact lay notions, and upon the availability of drugs to prescribe. What etiological research there was tended to be done by neuropsychologists and neurologists instead.

The DSM is a case of a classification that is moving away from Theory rather than to it, largely because it is not an attempt at a natural system, but one of convention and operational use. However, the majority of those who employ it seem to think it is a natural scheme. This may retard the progress of psychiatry, as Dom Murphy thinks:

> classification can draw on causal discrimination in the absence of causal understanding. And it can use causal discrimination as a source of hypotheses. If we have good reason to believe [through this classification] that two syndromes depend on different pathologies, then we can orient research around finding out what they are.... The system of classification in the DSM is incoherent, heterogeneous, and provincial. It is incoherent in that it rests on a theory about the taxa of interest that requires symptoms to be expressions of underlying causes whilst at the same time it prohibits mention of these underlying causes in the taxonomy. It is heterogeneous in

that it does not classify like with like at appropriate levels of explanation. And it is provincial in that it is cut off from much relevant inquiry. These complaints...reflect worries about the current state of biological psychiatry as a whole, since DSM-IV-TR is its flagship.[19]

With the approval of the DSM-V in 2012, critics noted in review that it was an amalgam of convenience and in some cases putative special interests.[20] Apart from a connection with the available treatments, not all of which are clinically or epidemiologically tested, there is a minimal change of emphasis upon etiology in refining the categories of the manual. Nevertheless, the mere having of a systematic classification generates substantial research programs, and the results of neurobiological research has input much of what etiology there is in psychiatry.[21]

Disambiguating the Theory-Dependence of Observation Thesis (TDOT)

For the past half century it has been largely agreed that one cannot observe without prior Theory. This is rarely explicated, however, and there seems to be some ambiguity in the claims made. So we will offer a preliminary taxonomy of the TDOT.

When Hanson introduced the claim it was this:

Weak TDOT: "There is a sense, then, in which seeing is a 'theory-laden' undertaking. Observation of x is shaped by prior knowledge of x."[22]

This is rather vague, and it remained vague when Kuhn took it up and defended it:

What a man sees depends both upon what he looks at and also upon what his previous visual-conceptual experience has taught him to see.[23]

The target of Hanson and Kuhn's discussions was the positivist claim that observations were framed in the language of "sense-data", which were pre-theoretic as experiences and in terms of language (the usual example used was "red here now", which is pretty far from actual scientific observations). And there is something to Hanson's claim, but it remains rather unlikely that Kuhn's claim that an Aristotelian and a Galilean will see a pendulum differently because they are "[p]racticing in different worlds". While we can say that prior experience modulates observation, it seems overwrought to say that one's "worldview" (a term to despise as meaningless) constructs our observations like this.

The strong version of TDOT that Kuhn asserted is this:

Strong TDOT: One can only perceive what one's worldview [global theoretical commitments] permits and requires.

This is extremely circumscriptive. And it is (rarely for a philosophical claim) empirically false, unless we reduce the notion of "theory" to "dispositions", in which case it is barely trivially true. Children have no worldviews (although they certainly have experience to guide them) and yet they learn to see, feel, and hear, etc. A recent study[24] has finally resolved, empirically, the issue of whether people have the capacity to perceive objects when they were born blind but acquire sight as adults (the Molyneux Question; answer: not unless they can correlate sight with touch and proprioception), which raises the question how children can possibly learn to recognize objects at all. It seems that infants have a disposition to treat motion and touching as a reward feedback, and so they move their limbs at random until they touch something. Is that "theory"? Only in the mind of a theory-obsessed philosopher could it be so considered.

Harold I. Brown[25] gives the best summary discussion of the TDOT. Brown distinguishes six different theses:

1. The items we perceive are already infected with material from the theories we accept.
2. Scientists ignore evidence that contradicts their favored theories.
3. Observations that are undertaken to evaluate a comprehensive theory presuppose that very theory in a way that prevents an objective test of that theory.
4. All scientifically significant observations assume some theories besides the theory being evaluated; it is always possible to protect a favored theory by challenging these auxiliary theories.
5. Which observations scientists undertake is determined by accepted theory.
6. Observation reports must be expressed in the language of the theory being tested if they are to be relevant to the evaluation of that theory.

Thesis 1 is undoubtedly true. Nobody begins an investigation *tabula rasa*, and having some theories, even if they are half-formed and vague, must influence at least some acts of observation. But it is unclear that this is necessarily the case or that the "infection" is so significant as to determine the outcome of the observation.

Thesis 2 is also likely to be occasionally true. That doesn't mean it is always true. This claim is rather trivial, if occasional, or false, if the claim is it that it always occurs.

Thesis 3 is the main subject of discussion in the period after Kuhn. If your theory makes you take observations in ways that beg the theoretical question, then the practice of science becomes an exercise in telling oneself stories for whatever reason. That way, postmodernism lies.

Thesis 4 is true, and forms the basis for Lakatos' claim that a research program has a protected core (a view we think is not true, but is contingently a fact about some disciplines or programs, and only a few). The Duhem–Quine hypothesis makes this point against falsification of theories, but again it claims too much. A truly perverse scientist can always retain a favored hypothesis (like creationism) by abandoning other theories (like, say, the whole of physics), but this won't fly in scientific arenas.

Thesis 5 is the crucial sense we are concerned with in this series. The point is not that prior experience and theoretical commitments influence observation, but that they determine what is observed. This determination leaves no room for untheoretical phenomena.

Thesis 6 is a claim about the language of the reports of observation. It basically means that even if you can perceive a phenomenon, you can't represent it except in the language of the theory. This thesis leaves no room for representations of observations except in terms of the meanings assigned by theory.

Of course two or more of these senses may be included in a particular claim for TDOT, but there is a general characterization we may give of it that leads to a better grip on the notion:

General TDOT: An observation report O is dependent upon a theory T.

We can focus on

1. the report and the language used
2. the dependency relation
3. the theory or theories and their scope, and degree of influence upon the observation report O.

If we focus on the report (thesis 6), then the argument might be that using the terms of the theory in some way means that theories are incommensurable because they do not denote the same parts of the world and so we cannot prefer one over another on empirical grounds. This extends to technical terms and mathematical variables. Moreover, we can usefully consider ways in which neurobiology modifies and processes signals. Kuhn and others thought this was crucial here. We do not. If we perceive, it is not because an image falls on our retina, but because the entire apparatus is working, from lens to visual cortex and beyond. The older view is like saying that a camera does not record accurately because it uses silver halide chemistry to indicate where photons fell, or in a digital camera, a CCD chip and internal processing and storage in RAW format.[26]

If we consider the dependency relation, a weak dependency (theory biases but doesn't determine observation) causes us only a little discomfort. For it to be problematic we need to have a strong claim, that is determines wholly (or mostly) what we can perceive and how. Only then will it exclude ordinary observation, naive observation, and phenomena that are not in the Bogen–Woodward definition: patterns in (theoretically determinate) data. If we allow that the dependency is partial, either because the theory is only slightly relevant or because it only biases our observations, then the way is clear for a kind of empiricism to both develop and establish theories more or less objectively. Therefore we really need to consider what the theories are on which observations depend, and in what manner.

Domains and theories in science

> If we examine some relatively sophisticated area of science at a particular stage of its development, we find that a certain body of information is, at that stage, taken to be an object for investigation.[27] [Dudley Shapere]

In this section we want to discuss the theoretical relativity of domains. If observation is hostage to Theory, the question immediately occurs: which Theory? One answer is that it is either the Theory of the domain in question, say, biology or immunology, or it is from a Theory external to the domain (Hacking's point about optics[28]). Theories are either of that domain or not, or perhaps the domain is part of a wider domain defined by a general Theory. How do we tell?

In part, this is made more difficult because it is not entirely clear either what a Theory is in the literature (there are a number of conceptions of theories), or what a domain is. The latter question used to be a major topic in discussions of science in the latter half of the nineteenth century. In the period since it seems to be discussed under the rubric of reduction of Theories, and so in the modern literature, a domain is defined by its Theory. Hence, on this account, biology is a domain, or not, because it has (or doesn't) a ruling Theory. Usually this is supposed to be Evolution (which is, in our view, at least six Theories, not one). Opponents of the evolutionary hegemony will offer Genetics, Development, or Ecology as defining global theories. Of course, if you do not have a Theory, but have a domain, then it gets interesting.

However, this is a debate over contingencies of scientific development. It matters not a whit to nature that we choose to isolate biology from, say, chemistry (talk to a biochemist about that) or from thermodynamics (again, there are biological thermodynamicists) and so on. It must not be thought that the disciplinary structure of our present science is itself a natural classification and carves nature at any significant joints. In the usual parlance, it is unclear whether they are natural kinds themselves. Yet, studies like Okada and Simon's[29] on collaborative discovery in a domain do presuppose that the domain matches up to some *sui generis* field of discovery and investigation.

If we presume that a domain is defined by Theory, at least when it is a nascent field, then it is hard to see how a field might evolve in the first place. But if we think the way the received view does, that a domain is a psychological, social or arbitrary field of enquiry, then it is hard to make out arguments about domain-specific theories, counter-reductionism, and the naturalness of a subject such as, for example, biology. If biology is simply what we prefer to identify, because of our prior conceptual dispositions (consider Gelman's argument for the innate essentialistic disposition of children regarding living things in her *Essential Child* [30]), no inferences other than about our psychology may be made about domains (and domains are indeed largely the grist for educational psychology, for example, Lawless and Kulikowich[31]).

So the bounding of natural domains by global Theory is a way to ensure that we can make inferences about objects within that domain. This leads rather directly to the theory-dependence of ontology (TDOO). We appear to have a domain conundrum:

If TDOT5 is correct, and TDOO is correct, then no new scientific domain may arise from a prior state of ignorance and lack of Theory regarding that domain.

The only way to resolve this is, in our opinion, to deny either TDOT5 or TDOO. We deny both, in the sense that they are not necessary for all domains (they may be true of some domains). But first, let us consider under what conditions a domain may develop.

For a domain to identify some phenomena that are natural, we may approach these phenomena in one of two ways. The phenomena may be derived from Theory that is external to the domain (Theory$_E$), which would mean the ontology of these phenomena, and the data on which they are based, is not an ontology/data suite from the domain in question. For example, atoms and the ordinary chemical kinds of abiotic chemistry are not biological entities even though they play a crucial role in explanations of the aqueous media of cells and interstitial fluids. So one may begin to examine chemical phenomena in biology without there being a biochemical Theory (yet). This permits us to iteratively refine our categories and observe phenomena in biochemistry to the point where they can be classified in terms of some kind of Theory independently of the assumed chemistry kinds.

If the Theory$_E$ is too distant from the data and phenomena of the domain itself (quantum mechanics relative to most biology, for example), then the kind terms in a domain without Theory cannot be bootstrapped this way. Of course, it may turn out as a contingent historical thesis that no domain has ever begun like this – though we think it has in several instances (taxonomy in biology and mineralogy being two cases, and arguably also in nosology, medical classification, and even in the case of the development of the periodic table, which did rely upon laboratory techniques of refining elements and weighing them, but lacked much other Theory[32]) – in which case we would treat this as a limiting case.

Moreover, it is likely that the notion of a domain is itself relative to background knowledge anyway: is mammalogy a domain independent of biology? What about protein chemistry? The nineteenth century is replete with examples of debating such questions (for example, Jevons[33]), leading to the view that it was, after all, merely a matter of convenience and librarianship.[34] Yet, we still debate whether or not biology is reducible to physics or has some irreducible core;[35] the domains are taken seriously in this debate.

The domain conundrum may be resolved if we accept external Theory and ontology, but if we have no recourse to this solution, what then? The answer is to understand that domains can evolve both culturally and cognitively-psychologically in order to reliably demarcate natural phenomena. Psychological evolution is basically about perceptual and cognitive evolution: we see objects that are there in the ordinary sense because if we did not, we tended not to pass on our cognitive and perceptual traits: "Creatures inveterately wrong in their inductions have a pathetic, but praiseworthy, tendency to die before reproducing their kind" as Quine wrote so eloquently but prescriptively.[36] So domains may begin through the construction of our evolved psychological traits, and then be refined. Moreover, a domain may begin as a social construct (for example herbalism in the Middle Ages), and then be refined by experience to become more natural, leading to Theory in the domain (Theory$_D$).

In these cases, Theory$_D$ does not determine what is observed initially, because there is no Theory$_D$. We begin with "ready-made" phenomena$_D$, but they are not handed to us on a plate by Nature. We surely have to construct them; only we do not need to already have observation languages and techniques unique to the domain under investigation. We can even construct the domains and be relatively sure they are more or less natural without a theory for that domain.

To finish, we offer this diagram, based on a metaphor of Papineau's.[37] Note that there are no hard and fast divisions between Theory$_D$ and observations$_D$, nor even (despite the sharp lines) between the domain itself and the external influences on observation. Theory and Observation are asymptotic ideals, possibly never to be reached.

What makes a domain historically appears to be the sequence of investigations that led to it being defined, and sociologically the political and other power relations that form theoretical traditions and research programs in institutions. But formally, there appears only to be epistemic boundaries drawn in order that a subject of investigation can be tractable given the techniques, computational power, and cognitive apprehension of the investigators. As Shapere said,

> in order that the area constituted by the related items be an area for investigation, there must be something problematic about it, something inadequate in our understanding of it. A domain … is not merely a body of related information; it is a body of related information about which there is a problem, well defined usually and raised on the basis of specific considerations.[38]

We would add: that the area must be tractable and addressable. The definition of a problem area for investigation is usually specified by it being located among relatively well-defined areas that have been addressed, from which this adjacent domain can be entered into. Shapere noted this, by setting out the characteristics of a domain, the fourth of which

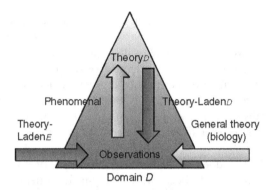

Figure 6.1   The relationship of Theory and observation in a specific biological domain, and the inputs from other domains and wider general knowledge

is that "Science is 'ready' to deal with the problem". This dialectic aspect to science is sometimes overlooked because of presentism and Whiggism, and the ever-present Theory-centrism that even Shapere exhibits in an otherwise marvelous essay.

We need not immediately presume that a domain like, say, soil science (see Chapter 4) is ontologically unique, nor that having a Theory in that domain will necessitate a metaphysics. It is enough to have natural phenomena to explain. Domains will shift at different stages of the historical development of a science. As do the entire sciences themselves.

Conclusion

We began this chapter by asking what the species concept referred to and whether or not it was necessarily a Theoretical concept, and hence either a unitary concept based upon a monistic Theory, or a plural concept based upon many Theories. We see now there is a third option: they are phenomenal objects that form the explananda for Theory in some cases. Given that there are many ways in which species is applied across the living world, and the lack of a viable rank or level of organization that species (and only species) apply to, we now can explain why biologists nevertheless continue to identify and strive to explain species.[39]

We have three conclusions: (1) The recognition that there are phenomenal objects in domains that might not be constrained by theories of that domain leads to the possibility of theory-independent classification. This suggests that (2) scientific theories might develop in a manner that is more like classical empiricism than usually thought. Finally, (3) phenomenal objects in a domain form the target of explanation by theory rather than necessarily being derived from it Duhem-style.

Notes

1. Swainson 1834, 105.
2. See Wilkins 2009a, 2009b, 2011.
3. Ingo Brigandt first made the claim that species are not theoretical objects (Brigandt 2003b), to Wilkins best knowledge.
4. Bogen and Woodward 1988, also implicit in van Fraassen 2008.
5. Massimi 2008a, 2008b, 2011.
6. Hacking 1983, 222f.
7. Hacking 1983.
8. Bradie 1986; Bradie and Harms 2004; Quine 1969b. See Lorenz and Cranach 1996, 18 for the a posteriori claim.
9. Wilkins 2009b, 38f and citations therein.
10. Blaxter 2004; Kress et al. 2005; Page, Choy, and Hughes 2005; Paul and Gregory 2005. For criticisms, see DeSalle, Egan, and Siddall 2005; Ebach and Holdrege 2005; Wheeler 2005; Will, Mishler, and Wheeler 2005; Will and Rubinoff 2004.
11. Winston 1999.
12. Van Fraassen 1980, 15.
13. Scerri 2007.
14. Scerri gives notice of at least four later "precursors", Emile Beguyer De Chancourtois, Gustavus Hinrichs, John Newlands and William Odling, who proposed some of the features later credited to Mendeleev.
15. Scerri 2007, 125, 150.
16. Scerri 2012 has argued that Weisberg 2007 is wrong to think Mendeleev was producing, inadvertently or otherwise, a model. He proposes that classification is a kind of "sideways explanations". This need not affect Weisberg's view of the importance of models (since he allowed that Mendeleev was not actually making models, but gave a hypothetical interpretation in which he was), but it is indicative of the lack of philosophical attention generally given to what Weisberg calls mere classifications.
17. Quoted in Kultgen 1958, 180.
18. This entire section is based upon the comprehensive research and discussion in Murphy 2006. Although Murphy's theoretical discussion of classification in chapters 9 and 10 is consonant with ours, we arrived at similar ideas independently and in distinct domains.
19. Murphy 2006, 323.
20. For example, the *Canadian Medical Association Journal* editorial, 10 December 2012 <http://www.cmaj.ca/site/earlyreleases/10dec12_the-perils-of-diagnostic-inflation.xhtml>. Child psychiatrist Claudia Gold refers to DSM-V as a "care rationing document" <http://claudiamgoldmd.blogspot.com/2012/12/a-relational-view-of-dsm-v-care.html>.
21. Kupfer and Regier 2011; Regier et al. 2009.
22. Hanson 1958, 19.

23. Kuhn 1996, 113 [1962].
24. Ostrovsky, Andalman, and Sinha 2006.
25. Brown 1996.
26. Cf. Katz 1983 on the picture theory of observation.
27. Shapere 1977, 518.
28. Shapere 1997, chapter 11.
29. Okada and Simon 1997.
30. Gelman 2003.
31. Lawless and Kulikowich 2006.
32. Scerri 2007.
33. Jevons 1878.
34. Richardson 1901.
35. Dupré 1993, 2006; Rosenberg 1994, 2006.
36. Quine 1953. See Griffiths and Wilkins 2013, Wilkins and Griffiths 2013.
37. Papineau 1979.
38. Papineau 1979, 521f.
39. Wilkins 2003, 2007a.

7
Radistics: A Neutral Terminology

Out of the trunk, the branches grow; out of them, the twigs.
[Herman Melville[1]]

Classification is a disputed territory in every science. Sometimes this is a professional matter, as when a term's meaning or application is disputed by competing schools or alliances. This, while it has more to do with professional authority and the dynamics of a discipline than natural kind terms, affects the way science is done more than one might suspect. How can we deal with the blossom of confusion over terminology and the "true" meanings of shared terms in natural classification? Allow us to play a little game here, and to define a neutral terminology for classification across all scholarly fields. We have chosen to call it Radistics, from the Greek root word for "branch": radix (ῥάδιξ), although the rest of the terminology is Latinate.[2] Its aim is to produce a schema into which the debates can be placed.

The use of terms in biological systematics

Classification is a general epistemic activity; it is not restricted to the natural sciences or any particular science. Nor is the practice of classification something that depends upon a particular philosophy of classification (or non-philosophy, if we take a Felsensteinian approach); it has commonalities across all fields and disciplines. However, much of the debate that surrounds classification has to do with the inexact agreement of terminology. For example, "class" itself has many different meanings, depending on the field in which it is being used. It can mean a defined (intensional) set, a type, some group or other, a universal, a property, or a taxon, or some mixture of all of these. Presently, fields such as systematics and biogeography, or soil science (pedology), for instance, borrow the same terms from other fields but give them different meanings. This leads to confusion and an endless debate that constitutes a sociological nightmare.

For example, terms like monophyly and its complement aphyly[3] (to include all non-monophyletic or unknown states such as paraphyly, polyphyly, polytomies and monotypic taxa), as well as homology have been defined to be "transformational" or "non-transformational (i.e., taxic)" by which is meant either a palaetiological or an abstract notion respectively. These definitions relate different concepts, which are gathered under a single term, and this has led to a conflict over which is the correct definition rather than which is the correct concept. Such terms unwittingly lead to classifiers talking past one another to the detriment of their discipline. For instance, anyone looking from the outside in will see a group of people with very different aims and goals engaging in debates that cannot be resolved with the limited terms available to them.

Therefore, we present a table of new terms in order to help repair the current communications breakdown. We offer "neutral" terms to indicate what aspects of classification each specialized technical term covers, so that we can cross-compare the different intents of each philosophical and scientific approach to classification. Although each term is new, the concept it represents is not, and we map existing specialized definitions to it. This terminology is presented only as a way to potentially resolve the conflict within systematics – it is not a replacement schema for any of the existing terminologies. If monophyly in the transformational sense and monophyly in the taxic sense play the same general roles in their respective approaches, it helps to have a placeholder category to identify the similarities and differences between them, rather than attempting to supplant one or the other based on Sharks and Jets conflict.

The rationale for Radistics, therefore, is to help classifiers in any discipline move on with their theory and methodology. The argumentation and definitions used throughout this book have employed existing terms. For instance, as we discussed, the term monophyly refers to several concepts, including but not restricted to "stem group", "all descendants and their most recent ancestor", "paraphyly" and "taxa more closely related to each other than they are to any other taxon". Regardless, the first and second definitions are still meaningful and used by most cladists and evolutionary taxonomists respectively.[4] Rather than argue which definition is "correct", matters would progress if we could relate it and other such terms to a new general category.

We begin with some basic terms (Table 7.1): *rade* for a group defined by relationships between taxa, shown as a *radogram* (any branching diagram that represents iconographicaly these relationships, topographically or including other means of identifying the degree of relationship such as variable branch lengths). A basic taxon comprises *units*, or *specimens*. How it is isolated and identified is a matter for each science.

Taxa are generally hierarchically arranged as subsets within larger inclusive sets. Some schemes do not exemplify this, while others (such as Linnaean taxonomy) have ranks of inclusive taxa. In the logical tradition that arose out of the Middle Ages, genera was a relative term, as was species, akin to the modern proper set/subset distinction that developed out of that tradition. There was no ranking as such apart from the summum genus and the infimae species, which were the largest set and smallest respectively. An infimae species comprises only individual objects, which we are calling units.

Hierarchical arrangements of taxa lend themselves to visual iconography in terms of trees. Although tree diagrams precede Darwin by a half century, arrangements of taxa into indented tables grouped using braces is older, and many have been misled into thinking these "tables" are tree diagrams, which is unfortunate given the connotations trees now have in a post-Darwin world, as transformational or historical sequences. As we have argued, the arrangement is the explanandum, and the historical account is the explanans. So to obviate this, we call these directed acyclic graphic representations of taxa radograms, which may be historical or may merely be taxic representations.

Aphyly in cladistics is the complement to monophyletic groups, as we said. This is a special case of a general problem of classification. There is and always has been a tendency to treat the remainder of a group, once a natural group has been excised from it, as a natural group in its own right. As we noted above, Aristotle criticized this approach as what came to be called privative groups, since what can be predicated of them does not uniquely identify them or mark them out from other groups (and note that Aristotle's discussion is in the context of his logic, not his natural science here, in *Posterior Analytics* I, 2–3. In his "Parts of Animals" however, he does not follow his own advice, drawing conclusions from animals being "bloodless"[5]). Since we need a term to connect units in taxa, we have chosen formism, as the diagnostic or essentialistic characters are here called *forms*, and hence the privative kind of taxa are in a state of *aformism*. Aformism is a more general term than aphyly, and covers what would be several cases in cladistics: unresolved taxa of all kinds, polyphyly, paraphyly, and cases of monotypy where an inclusive

Table 7.1 Equivalence table between some terms used in several sciences and Radistics including a list of definitions for radistic terms

Radistic Definition	Radistics	Logical grouping	Linnaean systematics	Cladistics	Pedology	Clouds
A group defined by a relationship between taxa	Rade	Class Kind Group	Natural group Taxon	Clade	[Taxon]	[Taxon]
A branching diagram depicting rades	Radogram	Directed acyclic graph	Table	Cladogram	N/A	N/A
A single specimen or object of classification	Specimen	Individual Element Member Prototype	Specimen Type specimen	Specimen Holotype, Paratype etc.	Pedon	Cloud
Basic kind of taxon	Infrataxon	Species Set, subset	Taxon Species	Clade Species	Soil	Category Species
Higher taxa	Supertaxon	Genus Proper set Superset	Genus Family Order Class...	Clade	Horizon Order Great Group Family Series...	Families/Étage Genera Species Varieties

	Form	Tuple?	Character essentialis / Trait / Morph	Homolog / Homology / Synapomorphy	Landscape
The parts of specimens that occur in other specimens as different forms and function	Form		Character essentialis / Trait / Morph	Homolog / Homology / Synapomorphy	Landscape
The relationship between two or more manifestations of the same formae	Formism	Relata / Differentiae / Intension		Three-item statement / Relationship / Relatedness / Ancestry	
The absence of formism between similar forms	Aformism	Partial set / Disjoint / Intersection		Homoplasy	
When taxa are more closely related to each other than they are to any other taxon	Holism	Shared nodes	Relationship	Relationship	
When taxa are unresolved	Aformism	Privative group / Complement		Polyphyly / Paraphyly / Monotypy / Polytomy	

taxon has only a single included taxon, thus making the character-states of the inclusive taxon identical to the included taxon, and thus indistinguishable. In effect a monotypic taxon has an indefinite description and is unknowable *qua* that taxon.

Sine qua non

At first, new terms are a nuisance, but with usage they become familiar, almost idiosyncratic to a particular field of study (in biological systematics, examples would be symplesiomorphies, apomorphies, etc.). This example highlights the need to enhance communication between specialists. Radistics uses new terms assigned to specific meanings in order to lower the chances of misunderstanding and help systematists enter a new phase of debate. This is not to say that everyone needs to agree or instantly understand the definitions of each new term. As systematists we only progress if we both understand these definitions and argue over their meanings and consequences. Currently we seem to be arguing over the terms rather than their meanings. For example, take the term cladistics:

> ... grouping by synapomorphy through the application of the parsimony criterion.[6]
>
> ... a method of classification that groups taxa hierarchically into nested sets and conventionally represents these relationships as a cladogram. See also phylogenetic systematics.[7]

What then are synapomorphies? Schuh and Brower[8] define synapomorphies as shared derived characters, which radistically are forms (i.e., equivalent to homologs) and not as relationships, that is, formisms. From a radistic interpretation, we may say that synapomorphies are simply statements of similarity (i.e., form) and not relationship (formism).

"Cladistics", being a poorly defined term, is synoptic of the entire issue. Scientists who operate under the rubric of "cladistics" not only use different methods, but also have radically different goals, motivations and definitions for fundamental terms like homology, homologs, monophyly and species. The debate above can now be clarified through radistic terms. Rather than state that some cladists are not interested in finding homology (interpreted as "character transformation") or monophyly (interpreted as "descendants and their most recent ancestor"), we may say that their intent is not to find formism or summology; that is relationship.

Presently Radistics is an association of terms with existing definitions to demonstrate a point – the communication breakdown in systematics. For those wanting to take Radistics a step further, namely to create a new philosophy of systematics, we issue a word of warning. Radistics does have elements of cladistics and departing from cladistics wholesale may exclude advocates from current debates (see Sharks and Jets, chapter 2). On the other hand, many other closely related fields have taken this step without detriment: numerical taxonomy, phylogeography,[9] molecular systematics[10] – becoming successful fields regardless of merit or demerit.

Tree-thinking

In recent thought, there has been a proposal that what separates evolutionary thinking, which is to say, Darwinian thinking, from earlier and now outmoded ways of thought, is something called "tree-thinking". As soon as it was proposed, it was under threat from an increasing emphasis on "lateral transfer" or "horizontal evolution". In this section we will first outline what tree-thinking is intended to compass, and then the challenges. Then we shall consider whether or not the pre-Darwinian views were actually tree thinking as described, and whether or not the modes of thought that are described as outmoded should be regarded as having either been supplanted by tree-thinking, and whether they are, in fact, in contradiction to it.

The term "tree-thinking" was proposed by O'Hara.[11] On his account, the previous view is class-thinking or essentialism, following the Essentialism Story line, but Darwin introduced a novel form of thought, which O'Hara deliberately modeled on "population thinking", which Mayr had asserted was Darwin's novelty.[12] Tree-thinking is the phylogenetic approach, best used in systematics, to explain "evolutionary events rather than the states of supposedly replicate species ... determining where the events occur on a phylogeny".[13] Tree-thinking, however, "does not necessarily entail knowing how phylogenies are inferred by practicing systematists. Anyone who has looked into phylogenetics from outside the field of evolutionary biology knows that it is complex and rapidly changing, replete with a dense statistical literature, impassioned philosophical debates, and an abundance of highly technical computer programs".[14] Tree-thinking is contrasted with "group thinking" which "equates 'systematics' with 'classification'".[15] However, O'Hara suggested that group thinking treats "members of a group as replicate instances" and dismissed the approach because it "breaks down for the fundamental reason that species are not independent replicates: they are parts of a connected tree of ancestry and descent".

Fundamentally, the novelty of Darwinian thinking according to O'Hara is to conceive of systematics as the discovery of lineages of ancestors and their descendants. Instead of thinking of species as distinct items unconnected with each other ("independent replicates" or "replicate instances"), as older thinkers often had, now they and the taxa they comprise are historical objects. They have beginnings and endings, as Hull and Ghiselin's individuality thesis noted for species themselves, and they have causal histories with other taxa. Group thinking is now abandoned. O'Hara formulated this modality of conception after a nasty debate and competition between pattern and process cladism. As we noted, almost from the beginning, pattern cladism was dismissed as "creationist", "typological", "essentialist" and outmoded thinking. O'Hara's treatment seemed to license this dismissal, and to evaluate the issues the way Mayr, and the process cladists independently, had, that their own preferred views were each the true successor to Darwin and modernity.[16] Of course, neither school of thought agreed upon the other being Darwin's descendants.

Similar arguments do not seem to have occurred outside of biology. As far back as Linnaeus' *Systema Naturae* in 1735, hierarchical classifications were applied to mineralogy for instance, but they seemed not to have taken root there, so to speak. Where classification schemes were indeed hierarchical outside of biology, this was a matter of definition and convenience rather than any attempt to capture the nature and etiology of the taxa. But even in biology, there have been attempts to denigrate the necessity and role of tree classifications.[17] Especially in the so-called "prokaryotes" (that is, the single celled organisms that are not eukaryotes), extensive cross taxonomic gene transfer is held to undermine the notion of a taxonomic tree, and the phylogenies that it implies. Similar arguments have been made against tree representations of language evolution.[18]

Groups and relationships

Not all taxonomic groups contain taxa that are more closely related to each other than they are to other taxa. Such groups are termed artificial. While this might be a problem for a systematist attempting to determine natural groupings, it does not necessarily mean that the group is unclassifiable. Many taxonomic keys are artificial, like the plant taxonomy of Linnaeus, but at the same time are incredibly useful for identifying taxa at all taxonomic levels. Artificial groups also have their uses outside of scientific classification. Folk taxonomies are rife with artificial classifications.

Take dinosaurs for example: every child knows what a *Brontosaurus* is, even though the name no longer is valid taxonomically (*Brontosaurus* has been synonymized by an older name *Apatosaurus*).[19] Regardless, Dinosaur is a term for discussing science and evolutionary biology with the general public. For example, we may tell a child holding a bird that he or she is actually holding a dinosaur. The effect is both instant and lasting: birds are seen as derived dinosaurs. The evolutionary biological message has been thoroughly conveyed with little scientific jargon. In biological taxonomy, however, the term "dinosaur" is as meaningful as "insectivore", "worm", "algae" or "invertebrate". Artificial classifications have their place in identifying and collecting known taxa.

Natural classifications, on the other hand, are about proposing relationships at the level of formisms. Here we can relate two or more taxa, which are either known and unknown, into a stable classification that will accommodate further closely related taxa without necessitating any later taxonomic changes. For example, all mammals have lactating glands and hair. Find taxa on Mars with these characteristics, they will relate more closely to mammals than to any other group. No changes are necessary to the taxonomy, that is to the diagnostic characteristics that delineate the group. Reptiles, however, are different. As more taxa are described, "reptiles" has needed changes to accommodate the new taxa. The reason is that "reptiles" contain taxa that share closer relationships to other groups (birds) than they do to other reptiles. Natural groups are predictive (all mammals have hair and lactating glands) while artificial groups are retrodictive, requiring redefinitions in order to accommodate new taxa. Within classification, groups can be artificial and natural, while relationships only point to natural groups.

Lateral transfer and horizontal connections

In 2009, *New Scientist* announced on the cover that "Darwin was Wrong".[20] According to the article, because lateral genetic transfer is known to frequently occur, the idea of an evolutionary tree (note: not a particular evolutionary tree but the very notion itself) has become moot. More sophisticated arguments for this have been made by philosophers[21] and biologists.[22] The problem with this is that unless there is a tree to compare against, the transfer of genes will appear just like sexual reproduction. In any case, Darwin's comments on phylogeny are restricted to plants and animals, where, despite hybridization, Darwin's idea is well founded. The real issue here is that such claims are based on a conflation of lineages – genetic and genomic with taxonomic.

There are three classes of genetic or molecular homological parallelisms: orthology, paralogy and xenology. Orthology is a genetic homology formed through a speciation event. Paralogy is a homology formed through a duplication of a gene (or gene product). Xenology is a homology formed by lateral transfer (Figure 7.1).

Without entering into a great deal of discussion, lateral or horizontal gene transfer results in xenology or "foreign genes",[23] which "is a form of homology (inferred common ancestry) in which the sequence (gene) homology is incongruent with that of the organism carrying the gene, and horizontal gene transfer or transfection is the assumed cause".[24] Colin Patterson equates xenology with parallelism in morphology, drastically downplaying its relevance in classification, other than causing

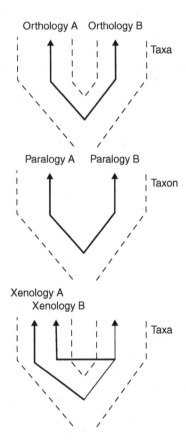

Figure 7.1   Varieties of molecular parallelisms

homoplasy (simple incongruence). Since it is not the task of systematics (but rather of phylogenetics and evolutionary biology) to resolve or interpret the specific causes of homoplasy, and indeed it is what systematics must perforce ignore, within a radistic framework, homoplasy (parallelism, xenology, paraxenology, etc.) is simply a non-formology.

These are homoplasies in phylogenetics. They are a case of convergence that is not informative about relationships and are therefore (except at the level of gene sequence) analogous relationships, based upon similarity metrics. For example, a paralogous gene might be a pseudogene (a gene that is not functional or expressed). These, as convergences, must be eliminated from gene trees when doing classification.[25]

Radistics as a language and iconography of classification

In setting up a neutral language for classification, we can usefully start to understand the real issues and clear up some of the confusions. As we noted in Chapter 3, the simplest classification is a pattern of three items or taxa that relates two items more closely to each other than either is to a third. All subsequent classifications are based upon this simple ordering of taxa. Note that when one cannot do this, the result is an unresolved classification, or in phylogenetics, a polytomy ("many cuts", which is to say we cannot specify a single ordering with the data available).

Now as we noted, tree diagrams are ubiquitous in classifications, and always have been. It has been regarded as an important aspect of classification since Aristotle that we should group things uniquely and unambiguously based upon positive properties held the items being classified [*Posterior Analytics, Categories*] and such arrangements naturally fall into a treelike structure. Initially, systematists (or their predecessors) used lists of taxa arranged using indentations to show subordination, and towards the end of the eighteenth century, the typographic convention began to include the use of braces (Figure 7.2, 7.3) to indicate higher ranking

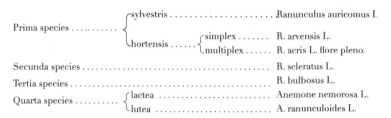

Figure 7.2 Klaus Fuchs' arrangements of species[26]

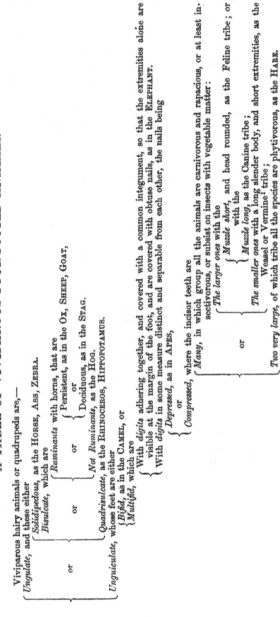

Figure 7.3 An early nineteenth century "table."[27]

groupings. About this time, too, the metaphor of a tree for systematic relationships came into being (see Chapter 3).

Tree-thinking has been called the most significant innovation of Darwinian theory,[28] and yet the metaphor, iconography and indeed the research problem that Darwin sets out to resolve with common descent preceded his work by some time. A tree structure is simply a way of representing nested classifications, and is formally equivalent to a number of other representations (Figure 7.4A) such as bracketed formulae, set diagrams, indented lists, legal numbering, binary trees (B-trees) in computer science, and so on. It is how the tree is interpreted

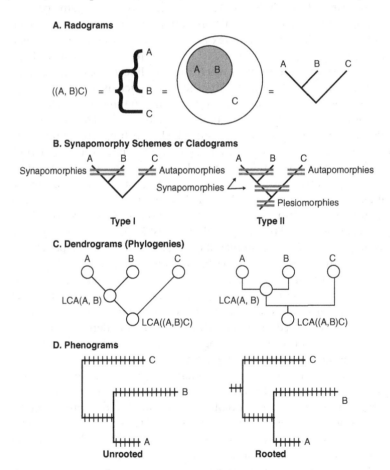

Figure 7.4 Tree diagrams. There are four general kinds of radogram. See text for discussion.

or read – that is, what it is taken to represent – that really matters. And this has been the subject of much confusion because it has not always or even often been kept clear what the iconography represents.

A basic Radogram (Figure 7.4A) is a simple hierarchical statement of the ordering of items or taxa, without any implication regarding the states of affairs that are being so ordered. Radograms can be represented as brackets, branching diagrams and Venn diagrams (or any other iconography that unambiguously represents the ordering). Terminal nodes are nothing more than the sum of the terminal branches and the base of the branching diagram (or first bracket), nothing more than one end of the statement. In evolutionary terms, for instance, radograms contain (as yet) no information about historical trajectories. Note that although this is unrooted, there is a vertex at the "base" of the diagram, which may be interpretable as a root. In fact it merely represents the less related taxon in the diagram. In cases where taxa are linearly related, this may represent a transformation, but in other cases, such as the relationships between soils or minerals, it merely represents the shared formae.

Cladograms (Figure 7.4B) are biological radograms. The synapomorphies within a cladogram indicate shared relationship, while autapomorphies indicate unique homologs. Plesiomorphies, however, are inferred. They are synapomorphies (shared characters) that are presumed to contain "primitive" or plesiomorphic character-states (homologs) with a linear transformation. Ebach et al.[29] have argued that the notion of transformations within cladograms are a posteriori inferences that have no bearing on overall typographical relationship. The inference of primitiveness is made by the observer of the tree, because the synapomorphy is "basal", that is, it lies lower within the hierarchy. Rather being interpreted as a necessary condition of the branching diagram, hierarchy is automatically inferred as a character-state transformation. The fact that taxon transformations are not observed[30] makes interpreting cladograms rather two-faced: homologs (the parts that are derived from taxa) can transform into other homologs, but taxa cannot. Cladograms are radograms, but not all radograms are cladograms.

Dendrograms (Figure 7.4C) are branching diagrams in which the internal nodes represent actual taxa. As such they are only one kind of cladogram interpretation. In biological systematics, the internal nodes are interpreted as last common ancestors (LCA's) of the subsequent branches of the tree. Such items or taxa are hypothetical, however, as when known taxa are represented in a cladogram, ancestral taxa as much as any others might be represented as "sister taxa", that is, as terminal nodes on the tree diagram. There are many different versions of dendrograms. Some of them indicate the stratigraphic order of the last known specimens (i.e., when taxa are thought to have gone extinct") in which case they are indicated by a dagger (or cross).

Phenograms (Figure 7.4D) rely upon the "degree of difference" established according to some similarity metric. Given the mosaic nature of evolution, it is highly unlikely that independently variable characters will all share the same degrees of difference, and so phenograms either do not represent genealogy (or more generally, etiological relationships), or they suppose that overall similarity metrics will converge upon them. We have argued against this in Chapter 4.

Wagner trees (cladogram type I, Figure 7.4C) represent the arrangements based upon shared character-states or synapomorphies. Wagner trees show synapomorphies and unique character-states (autapomorphies), and "primitive" (plesiomorphy) character-states. While synapomorphies and derived character-states are based on little inference (all homologs are derived in some fashion and shared characteristics are merely conditions of a node), plesiomorphies are inferred by an observer from a notion of character-state transformation. That transformation series is considered to be a homology, hence shared apomorphies (Synapomorphy) + shared plesiomorphies (Symplesiomorphy) = Homology. Another important characteristic of Wagner Trees is the notion of reversals. The notion of reversals and the ability to observe them on a tree is a defining characteristic of Wagner Parsimony.

The way reversals are inferred is by taking the hierarchy within a branching diagram to mean a transformations series. Conceptually, this is problematic, as the structure of a branching diagram is taken to be evidence for transformation. A reversal is inferred when a homolog is allowed (by the computer program) to form another grouping inside itself i.e., {0,{1,0}}. Rather than rejecting this as a flaw in the program or as a homoplasy, it is interpreted as a reversal. Overall, this means that transformations are only inferred if the observer believes that hierarchy and poor grouping procedures are phylogenetically meaningful. However, the logic behind reversals is that character-states may be lost within a group. Consider eye loss in cave fishes. This loss is interpreted from an absence (generally a plesiomorphy) on a Wagner Tree. This means that Wagner Trees may contain clades based on absences (reversals). Grouping on absences usually results in homoplasies within unparsimonious trees (a reversal creates an extra step).[31]

Statements of relationships (cladograms type II, Figure 7.4C), however, are constructed based on the relationship of two taxa based on a third, such as "my two cats Vesper and Purley are more closely related to each other than they are to me", hence {{Vesper, Purley}, Me}. Simple statements of relationship contain no plesiomorphies, since these are assumed, and are untestable and unobservable in the radogram.

Statements of relationship can also be made at the character-state level. We may say that a human forelimb is more closely related to a bat's fore-limb (namely a wing) than it is to a dolphin's forelimb (namely a fin). Since statements of relationship are about a hypothesis of relationship rather than character-state transformation, plesiomorphies are both redundant and not observable (can anyone see a primitive forelimb?). Apomorphies are naturally assumed to be real, as everything is derived and observable. This is also true of plesiomorphies, which by definition must also be an ancestral form, thus making them apomorphic at some stage.

Statements of relationships are often in themselves assumptions of common descent and are often confused with phenograms due to their lack of transformational series of character-states. The "classic" tree in biology is of course "the" evolutionary tree. What an evolutionary tree means, however, is hard to determine. It must at the least involve specia-tion events, and so its overall topology is genealogical, but most repre-sentations of evolutionary trees until the 1970s also included subtle and overt hints about the "degree of difference" in evolutionary events, or clustered groups on similarity grounds (not, it turned out, actual metrics, but some kind of subjective impression). The virtue of the numerical taxonomy movement ("phenetics") was that it made the notion of similarity precise. Its vice was that it applied this metric to any and all characters, irrespective of whether they were homologies or homopla-sies, due to acceptance of the operationalist view of the philosophy of science and a desire to be "theory-free" and "objective".

Monophyly: A term with many definitions[32]

The term monophyly was coined by Ernst Haeckel in the nineteenth century, but he meant by it roughly a joint notion of shared ancestry and degree of difference, which we are calling a hybrid classification. When Hennig appropriated the term, initially he retained also the ambi-guity of Haeckel's formulation, having two definitions. One was what has come to be the key concept of naturalness in cladistics: the group that includes the last common ancestor of two or more species, and all the descendants of that ancestor. However in his Grundzüge[33] he also mentioned another definition which included the descendants and not the ancestor. This was noted by Brundin in the work that made phyloge-netic systematics widely known.[34]

In 1971, Peter Ashlock took issue with the burgeoning consensus, noting that "[s]ystematists have long used the term monophyly and have felt sure they knew what was meant when the used it".[35] Ashlock referred to the cladistic definition as holophyly, a form of constrained monophyly, rather than the definition including the stem ancestor. Ashlock considered both holophyly and paraphyly as forms of monophyly.

However, there is another definition, a taxon that is more closely related to another taxon than it is to anything else. While Ashlock, and many cladists rarely refer to this definition, it too is termed monophyly (and "kinship" by Hennig). The latter definition is unique, as it is a statement of relationship that is independent of a phylogenetic inference. Reciprocal monophyly however, occurs when a gene lineage shares a more recent common ancestor than to any other lineage. While the definition of reciprocal monophyly is derived from Hennig's usage, it explicitly refers to gene lineages (tokogenesis) and not taxonomic lineages (phylogeny).[36]

Radistics redux

Biological classification is a mess of contradictory definitions of the same terms. Many, like Ashlock,[37] have attempted to redefine or at least reorganize these definitions into appropriate existing terms. While these attempts are helpful, they end up creating more terms that are ironically immediately rejected as confusing. Scientists it seems are more content battling over definition than terms. Hennig's definition of monophyly,[38] for example, is used in preference to Mayr's and Haeckel's. However, molecular systematists have created their own concept (reciprocal monophyly)[39], cleverly adding "reciprocal" in order to differentiate it from the term used by those working above the species level. While this reduces the level of confusion, it does introduce a new definition more similar to Haeckel's usage. Homology is in similar trouble, currently having undergone a change to its terminology from the Hennigan "synapomorphy" to the numerical "synapomorphy plus symplesiomorphy". Again, these changing and conflicting definitions create chaos within the literature, with systematists regularly talking past each other in great fervor. While this does not bother the practitioners, it does create problems for the philosopher of science wanting to compare these definitions without seeming to take sides. The purpose of Radistics, to introduce a whole new set a neutral terminology, has the potential to help philosophers and historians bypass needless argumentation in order to make a poignant claim without facing the wrath of systematists.

Notes

1. Moby Dick, chapter 63.
2. If we had chosen to name this in a Latinate manner, we would have had to call it "palaetiological systematics", which is nearly as awkward as "phylogenetic systematics" (also a Greek-Latin hybrid). So Radistics it is. Purists will have to cope.
3. A term coined by Ebach and Williams 2010.
4. Williams, Ebach, and Nelson 2008.
5. See Lennox 2002, 294.
6. Schuh and Brower 2009, 259.
7. Kitching et al. 1998, 202.
8. Schuh and Brower 2009, 268.
9. Avise 2009.
10. Hillis et al. 1996.
11. O'Hara 1991, 1993, 1996.
12. Hull 1994; Levit and Meister 2006; O'Hara 1997; Sober 1980.
13. O'Hara 1997, 325.
14. Baum, Smith, and Donovan 2005, 979.
15. O'Hara 1997, 324.
16. See, for example, Patterson's description of the furore in the Natural History Museum and in the pages of Nature Patterson 1999. Also see Hull 1988; Williams and Ebach in press. The "official" perspective is expressed in the widely-read textbook, Ridley 1986.
17. Makarenkov and Legendre 2004; O'Malley and Boucher 2005; O'Malley 2010; Ochman, Lawrence, and Groisman 2000; Ochman, Lerat, and Daubin 2005; Ragan 2001; Ragan, McInerney, and Lake 2009.
18. For language phylogenetics, see Bryant, Filimon, and Gray 2005; Gray and Atkinson 2003; Hoenigswald 1987; Mufwene 2001; O'Hara 1996; Wilkins 2002; contra: Stamos 2002.
19. Gould 1991 bemoans this, to no avail.
20. Edition January 21, 2009, by Graham Lawson.
21. O'Malley and Dupré 2007.
22. Brüssow 2009, Doolittle 2009.
23. Gray and Fitch 1983.
24. Patterson 1988a, 612.
25. Hillis 1994.
26. Redrawn from Sprague and Nelmes 1928/1931, 557.
27. Owen 1859, 5.
28. O'Hara 1991, 1996.
29. Ebach, Williams, and Vanderlaan 2013.
30. Except in the case of linear geographical dispersal in the Progression Rule.
31. A great deal of discussion has taken place in the pages of *Cladistics* regarding the nature of reversals and homology. We refer the reader to Brower and de Pinna 2012 and Williams and Ebach 2012. According to Laet and Smets 1998, the smallest unit of relationship is required to include a reversal is a 4-item statement, namely {0{1{0,0}}}. Adding a reversal creates an extra step, making a 4-item statement less parsimonious when compared to a 3-item statement.

32. We are deeply indebted to Tegan A. Vanderlaan for her thesis work on this. She is preparing a manuscript of the definitions and their history. We do not claim originality here: this is all Tegan's.
33. Hennig 1950.
34. Brundin 1966.
35. Ashlock 1971, 63.
36. Knowles and Carstens 2007.
37. Ashlock 1971.
38. Hennig 1950.
39. "Reciprocal monophyly, in which case all alleles within each sister taxon are genealogically closer to one another than to any heterospecific alleles", Avise 1994, 128.

8
Worth the Knowing

I'm looking through you/You're not the same.[1]

What classifications contribute to the inferential process in science is that they allow us to locate the mass of data points observed without Theory in a broader pattern, and they guide Theory-building. Classification is not, in and of itself, Theory-building; nor is it free of Theory when Theory is available. However, if we have no Theory, or the Theory is contested, then we should recognize that a classification scheme is a statement of what we do know, and rest easy in our ignorance of what we do not.

Mommsen and the Etruscans

In 1862, in the first volume of his monumental *History of Rome*, Theodor Mommsen famously wrote

> It is equally difficult to determine from what quarter the Etruscans migrated into Italy; nor is much lost through our inability to answer the question, for this migration belonged at any rate to the infancy of the people, and their historical development began and ended in Italy. No question, however, has been handled with greater zeal than this, in accordance with the principle which induces antiquaries especially to inquire into what is neither capable of being known nor worth the knowing...[2]

"Neither capable of being known nor worth the knowing" sets up the critical issues of classification and the paletiological (historical) sciences in general. Some deny that aspects of the past are capable of being known. Others deny they are worth the knowing. Some assert that no matter what the evidence may show, we can know the past processes that lead to the observed outcomes, through theory and models. Ironically, Mommsen's principled objections have recently been resolved in ways that illuminate our topic. Two studies – one that compared Etruscan (modern Tuscan) DNA to that of haplotypes in the Middle East, and another that did the same thing for the genotypes of the cattle of both regions – have tended to support the long-deprecated claim by Herodotus and Strabo that they came from Lydia, now Anatolia in central Turkey.[3] So it was, indeed capable of being known (at least to the degree of certainty conferred by such studies). Was it worth the knowing?

A simple statement that the Etruscans arrived around 1200 BCE from central Turkey is only interesting if the question of interest is where their center of origins was. That is a very slight topic, however, and once known, it can be filed away under "miscellany" and cited occasionally. History is full of facts that are slightly interesting once uncovered. No historian, however, would consider their subject to be principally engaged in finding out only such facts. The facts, the historical events and phenomena, must be placed in a wider context and employed in explanations that are much greater than the factual content. "Etruscans came from Anatolia" has little significance until it is placed in the context of the migrations caused by, among other events, the Phrygian invasions of the region, and the general disaporas of the time out of which the more recent dispersals can be seen to have emerged. This period was in fact a period of great migration and displacement, the like of which has not been seen until quite modern times. Against that comparative background, and the patterns it displays for the region and period, the Etruscan origin is quite important. It shows that when compared to the broader concerns of the field of investigation, facts uncovered by research form part of a more general phenomenal pattern. By means of this a discipline advances: each new pattern raises questions that call for explanations, and each verified phenomenon or fact gives a new pattern.

Historical phylogenetics has always presumed that the task of phylogenies is to uncover, and as Sober put it, reconstruct, the past. Pattern cladism has denied classification that role – and contrary to the usual characterization, pattern cladism has not rejected the possibility of reconstructing past sequences, only the flat claim that a classification is either ipso facto such a reconstruction or that one must have some model of the way living things evolve in order to classify. In the context of the views presented in this book, these issues are somewhat sharper. This is a matter of empiricism, theory-dependence, and most importantly the temporal ("erotetic") relativity of what counts as a fact or a theory. What cannot be known and is not worth knowing at one time can be known, empirically, with new techniques at another and contribute to further, comparative, research questions. The concern is that we do not replace observation with theory and think that we have made some progress. Science is founded upon empirical observations, no matter how these are tied up with local and cross-disciplinary theoretical commitments or stances. Once we abandon this aspect of science, whether because we are unreflective or because we are of the philosophical view that everything is derived from theory, science becomes little more than a matter of worldviews and epistemic statements of faith.

No scientist acts as if everything is derived from theory, not even those who assert this claim. Peter Medawar, in his declaration for Popperian falsificationism, wrote[4]

> Innocent, unbiased observation is a myth: "experience is itself a species of knowledge which involves understanding," said Kant. What we take to be evidence of the senses must itself be the subject of critical scrutiny. Even the fundamental principle of empiricism is open to question, for not all knowledge can be traced back to an origin in the senses. We inherit some kinds of information.

And yet, it would seem that what Medawar won his Nobel medal for involved meticulous empirical and experimental work on immunity.[5] It even involved inductive inference despite his rejection of induction in this later book. Of course, inductivism and empiricism do not lead research, but they are often important. The Kantian phenomenon/ noumenon distinction, like the equal and opposite strict empiricism and inductivism of those Medawar opposes, is incomplete and overly absolute. Night shades into day.

The role, or more correctly a role, of classification is to provide both the basis for discovery based upon patterns, which even if it is not induction is ampliative reasoning based upon as little theoretical foundation as possible, and to set up the contrasts for future research questions. We could not investigate whale lactation until we classified them as mammals.

Our main problem is that without classification, scientific inference can be based upon ignorance, piled upon ignorance. We reconstruct the history of a group like the Etruscans with many prior assumptions – about breeding rates, local convergences of gene sequences, migrations, the least complex presumptions about how genes distribute over time, and so forth. All these are reasonable assumptions, but they are not knowledge. They are hypotheses. Whether or not one's philosophy of science involves simple empiricism or a hypothetical-deductive model of explanation, or one adopts a more sophisticated Bayesian or Likelihoodist view,[6] evidence must play a role in deciding between hypotheses, not more hypotheses.

Claude Bernard, who effectively invented modern physiology and who along with Darwin, Mendel and Pasteur is one of the greats of the life sciences of the nineteenth century, used to rail against this reliance upon hypotheses as evidence, and instead embraced ignorance as a necessary and virtuous aspect of science.

> I live in a state of ignorance; therein is my philosophy. I have tranquillity of ignorance and faith in science. Others cannot live without faith, without belief, without theories that explain everything. These, I do without. I sleep on the pillow of ignorance.[7]

Ignorance is a frightening state to be in, especially when one wishes to know something about a topic. But it can be a pillow on which to sleep until we have the information to know. We always lack knowledge about something: this is why we do science in the first place. We do not solve problems in science by taking guesses and piling more guesses upon them and calling it knowledge. We should not be afraid to say that we do not know.

Some things may be forever hidden to us: it is unlikely we will ever have a great degree of certainty about most ancestors, for instance. We can tell stories, and make guesses, but in the end they remain hypotheses, and we cannot rightly test new hypotheses on the basis of previous hypotheses. Since proof is so often missing in science, all we can do is test for coherence between our most admired hypotheses in the absence of observational data.

For example, it is likely, given what we think we know, that *Homo erectus* is an ancestor of *Homo sapiens*, but when, where and what comes in between us and them is moot unless we get clear information, and clear information evaporates over time and space. Science is a fallibilistic enterprise, and no amount of polishing the turd can disguise this. Instead, it is best that we accept this and do our best. We are restrained by our prior hypotheses and likelihoods.

Observational (as well as experimental) classification provides an anchor point for hypotheses. Contrary to the Popperian views of some systematists, a classification is not a hypothesis, of ancestry or anything else. Once you have a natural classification, you can explain it with a hypothesis or model, and you can test that hypothesis with more observational classifications, but the hypothesis itself is not data. It is what it is: hypothetical. Nothing but trouble can come from confusing the explanation with the explanans, as we saw regarding the concept of homology.

So natural classification is an under-appreciated and understudied aspect of science. We hope that this will change.

Notes

1. The Beatles, 'I'm Looking Through You', 1965, From the *Rubber Soul* album, EMI Records.
2. Mommsen 1862, chapter 9.
3. Achilli et al. 2007, Pellecchia et al. 2007. Archeological, linguistic and documentary evidence is discussed by Beekes 2003, who suggests that the Etruscans came from a region to the north of the modern Lydians, and that Herodotus and Strabo, writing some 800 years later, were confused by this (see also Mahoney 2008).
4. Medawar 1969, 28.
5. Billingham, Brent, and Medawar 1954, 1956; Billingham et al. 1954.
6. For example, the Likelihoodism of Sober in his recent book (Sober 2008). Again, we must point out that an underlying philosophy of science presented as "parsimony", "likelihood" or Bayesianism, is not automatically acceptance of the phylogenetic or systematic analytic techniques or algorithms that go by those names.
7. From Nelson and Ladiges 2009. They write,"The quote is from a note ('*un autre papier inédit*'), said to be written in a juvenile hand [Chevalier 1937, 50; Mauriac 1940, 149, and Cotard 1944, 53, who use '*théologie*' in place of 'théorie']".

Bibliography

1908. "Discussion of the Species Question." *The American Naturalist* 42 (496): 272–281.

Achilli, Alessandro, Anna Olivieri, Maria Pala, Ene Metspalu, Simona Fornarino, Vincenza Battaglia, Matteo Accetturo, Ildus Kutuev, Elsa Khusnutdinova, Erwan Pennarun, Nicoletta Cerutti, Cornelia Di Gaetano, Francesca Crobu, Domenico Palli, Giuseppe Matullo, A. Silvana Santachiara-Benerecetti, L. Luca Cavalli-Sforza, Ornella Semino, Richard Villems, Hans-Jürgen Bandelt, Alberto Piazza, and Antonio Torroni. 2007. "Mitochondrial DNA Variation of Modern Tuscans Supports the Near Eastern Origin of Etruscans." *American Journal of Human Genetics* 80 (4): 759–768.

Agassiz, Louis. 1860. "[Review of] On the Origin of species." *American Journal of Science and Arts* (Ser. 2) 30: 142–154.

Anderson, Erik. 2005. "How General is Generalized Scientific Essentialism?" *Synthese* 144 (3): 373–379.

Anderson, Lorin. 1976. "Charles Bonnet's Taxonomy and the Chain of Being." *Journal of the History of Ideas* 37 (1): 45–58.

Archibald, J. 2009. "Edward Hitchcock's Pre-Darwinian (1840) 'Tree of Life'." *Journal of the History of Biology* 42 (3): 561–592.

Aristotle. 1997. *Topics: Books I and VIII*, with excerpts from related texts, Clarendon Aristotle series. Translated by Robin Smith. Oxford: Oxford University Press.

Aristotle. 1995. *The Complete Works of Aristotle*. Revised Oxford translation. Edited by Jonathan Barnes, Bollingen series; 71: 2. Princeton, N.J.: Princeton University Press.

Armstrong, D.M. 1978. *Universals and Scientific Realism*. Cambridge: Cambridge University Press.

Ashlock, Peter D. 1979. "An Evolutionary Systematist's View of Classification." *Systematic Zoology* 28 (4): 441–450.

Ashlock, Peter D. 1972. "Monophyly Again." *Systematic Zoology* 21 (4): 430–438.

Ashlock, Peter D. 1971. "Monophyly and Associated Terms." *Systematic Zoology* 20 (1): 63–69.

Assis, Leandro, and Ingo Brigandt. 2009. "Homology: Homeostatic Property Cluster Kinds in Systematics and Evolution." *Evolutionary Biology* 36 (2): 248–255.

Assis, Leandro C.S. 2013. "Are Homology and Synapomorphy the Same or Different?" *Cladistics* 29 (1): 7–9.

Atran, Scott. 1990. *The Cognitive Foundations of Natural History*. New York: Cambridge University Press.

Avise, John C. 2009. "Phylogeography: retrospect and prospect." *Journal of Biogeography* 36 (1): 3–15.

Avise, John C. 1994. *Molecular Markers: Natural History and Evolution*. New York: Chapman and Hall.

Ayala, Francisco José. 1988. "Can 'Progress' be Defined as a Biological Concept?" In *Evolutionary Progress*, edited by Matthew H. Nitecki, 75–96. Chicago: University of Chicago Press.

Babich, Babette E. 2003. "Kuhn's Paradigm as a Parable for the Cold War: Incommensurability and its Discontents from Fuller's Tale of Harvard to Fleck's Unsung Lvov." *Social Epistemology* 17 (2–3): 99–109.

Bailey, Kenneth D. 1994. *Typologies and Taxonomies: An Introduction to Classification Techniques, Quantitative Applications in the Social Sciences.* Thousand Oaks, CA: Sage Publications.

Barker, Matthew J. 2010. "Specious Intrinsicalism." *Philosophy of Science* 77 (1): 73–91.

Basinski, Jan Jerzy. 1959. "The Russian Approach to Soil Classification and its Recent Development." *Journal of Soil Science* 10 (1): 14–26.

Baum, David A. 2009. "Species as Ranked Taxa." *Systematic Biology* 58 (1): 74–86.

Baum, David. 2008. "Reading a Phylogenetic Tree: The Meaning of Monophyletic Groups." *Nature Education* 1 (1).

Baum, David A., Stacey DeWitt Smith, and Samuel S.S. Donovan. 2005. "The Tree-Thinking Challenge." *Science* 310 (5750): 979–980.

Beekes, R.S.P. 2003. *The Origin of the Etruscans.* Amsterdam: Koninklijke Nederlandse Akademie van Wetenschappen.

Belon, Pierre. 1555. *L'histoire de la nature des oyseaux, avec leurs descriptions, & naïfs portraicts retirez du naturel: escrite en sept livres.* Paris: G. Cauellat.

Beltran, M., C.D. Jiggins, V. Bull, M. Linares, J. Mallet, W.O. McMillan, and E. Bermingham. 2002. "Phylogenetic Discordance at the Species Boundary: Comparative Gene Genealogies among Rapidly Radiating Heliconius Butterflies." *Molecular Biology and Evolution* 19 (12): 2176–2190.

Bentham, George. 1827. *An Outline of a New System of Logic. With a Critical Examination of Dr. Whately's "Elements of Logic".* London: Hunt and Clark.

Besson, L. 1921. "La classification détaillée des nauges en usage à l'Observatoire de Montsouris." In *Annales des Services Techniques d'Hygiène de la Ville de Paris.* Paris.

Billingham, R.E., L. Brent, and P.B. Medawar. 1956. "Quantitative Studies on Tissue Transplantation Immunity. III. Actively Acquired Tolerance." *Philosophical Transactions of the Royal Society of London Series B: Biological Sciences* 239 (666): 357–414.

Billingham, R.E., L. Brent, and P.B. Medawar. 1954. "Quantitative Studies on Tissue Transplantation Immunity. II. The Origin, Strength and Duration of Actively and Adoptively Acquired Immunity." *Proceedings of the Royal Society of London Series B: Biological Sciences* 143 (910): 58–80.

Billingham, R.E., L. Brent, P.B. Medawar, and E.M. Sparrow. 1954. "Quantitative studies on tissue transplantation immunity. I. the Survival Times of Skin Homografts Exchanged between Members of Different Inbred Strains of Mice." *Proceedings of the Royal Society of London Series B: Biological Sciences* 143 (910): 43.

Bird, Alexander. 2009. "Essences and Natural Kinds." In *Routledge Companion to Metaphysics*, edited by Robin Le Poidevin, Peter Simons, Andrew McGonigal and Ross Cameron, 497–506. Abingdon: Routledge.

Bishop, Christopher M. 1995. *Neural Networks for Pattern Recognition.* Oxford: Oxford University Press.

Blaxter, Mark L. 2004. "The Promise of a DNA Taxonomy." *Philosophical Transactions: Biological Sciences* 359 (1444): 669–679.

Blaxter, Mark, Jenna Mann, Tom Chapman, Fran Thomas, Claire Whitton, Robin Floyd, and Eyualem Abebe. 2005. "Defining Operational Taxonomic Units using DNA Barcode Data." *Philosophical Transactions: Biological Sciences* 360 (1462): 1935–1943.

Bogen, James, and James Woodward. 1988. "Saving the phenomena." *The Philosophical Review* 67 (3):303–52.

Bonnet, Charles. 1745. *Traité d'Insectologie ou observations sur les Pucerons*. Vol. 2. Paris: Durand.

Borgmeier, Thomas. 1957. "Basic Questions of Systematics." *Systematic Zoology* 6 (2): 53–69.

Boyd, Richard. 2010. "Homeostasis, Higher Taxa, and Monophyly." *Philosophy of Science* 77 (5): 686–701.

Boyd, Richard. 1991. "Realism, Anti-Foundationalism and the Enthusiasm for Natural Kinds." *Philosophical Studies* 61 (1–2): 127–148.

Boyd, Richard. 1999a. "Homeostasis, Species, and Higher Taxa." In *Species, New Interdisciplinary Essays*, edited by R.A. Wilson, 141–186. Cambridge, MA: Bradford/MIT Press.

Boyd, Richard. 1999b. "Kinds, Complexity and Multiple realization." *Philosophical Studies* 95 (1–2): 67.

Braddon-Mitchell, David, and Robert Nola. 2009. *Conceptual Analysis and Philosophical Naturalism*. Cambridge, MA: MIT Press.

Bradie, Michael. 1986. "Assessing Evolutionary Epistemology." *Biology and Philosophy* 1: 401–459.

Bradie, Michael, and William Harms. 2004. "Evolutionary Epistemology." In *The Stanford Encyclopedia of Philosophy*, edited by Edward N. Zalta, <http://plato. stanford.edu/archives/spr2004/entries/epistemology-evolutionary/>. Stanford University.

Brady, R.H. 1982. "Theoretical Issues and 'Pattern Cladists.'" *Systematic Zoology* 31: 286–291.

Brady, R.H. (1972). *Towards a Common Morphology for Aesthetics and Natural Science: A Study of Goethe's Empiricism*. SUNY PhD Thesis, 1–323.

Brigandt, Ingo. 2009. "Natural Kinds in Evolution and Systematics: Metaphysical and Epistemological Considerations." *Acta Biotheoretica* 57 (1): 77–97.

Brigandt, Ingo. 2007. "Typology Now: Homology and Developmental Constraints Explain Evolvability." *Biology and Philosophy* 22 (5): 709–725.

Brigandt, Ingo. 2003a. "Homology in Comparative, Molecular, and Evolutionary Developmental Biology: The Radiation of a Concept." *Journal of Experimental Zoology Part B: Molecular and Developmental Evolution* 299B (1): 9–17.

Brigandt, Ingo. 2003b. "Species Pluralism does not imply Species Eliminativism." *Philosophy of Science* 70 (5): 1305–1316.

Brigandt, Ingo. 2002. "Homology and the Origin of Correspondence." *Biology and Philosophy* 17 (3): 389–407.

Brigandt, Ingo, and Leandro Assis. 2009. "Homology: Homeostatic Property Cluster Kinds in Systematics and Evolution." *Evolutionary Biology* 36: 248–255.

Brigandt, Ingo, and Paul Griffiths. 2007. "The Importance of Homology for Biology and Philosophy." *Biology and Philosophy* 22 (5): 633–641.

Brower, Andrew V.Z., and Mario C.C. de Pinna. 2012. "Homology and Errors." *Cladistics* 28 (5): 529–538.

Brown, Harold I. 1996. "The Methodological Roles of Theory in Science." In *The Scientific Nature of Geomorphology: Proceedings of the 27th Binghamton Symposium in Geomorphology*, held 27–29 September, 1996, edited by Bruce L. Rhoads and Colin E. Thorn, 3–20. Hoboken NJ: Wiley.

Brundin, Lars Zakarias. 1966. *Transantarctic Relationships and their Significance, as Evidenced by Chironomid Midges. With a Monograph of the Subfamilies Podonominae and Aphroteniinae and the austral Heptagyiae, etc.* 472. pl. 30. Almqvist and Wiksell: Stockholm.

Brundin, Lars Zakarias. 1972a. "Evolution, Causal Biology, and Classification." *Zoologica Scripta* 1 (3): 107–120.

Brundin, Lars Zakarias. 1972b. "Phylogenetics and Biogeography." *Systematic Zoology* 21 (1): 69–79.

Brüssow, Harald. 2009. "The Not so Universal Tree of Life or the Place of Viruses in the Living World." *Philosophical Transactions of the Royal Society B: Biological Sciences* 364 (1527): 2263–2274.

Bryant, D., F. Filimon, and R.D. Gray. 2005. "Untangling Our Past: Languages, Trees, Splits and Networks." In *The Evolution of Cultural Diversity: A Phylogenetic Approach*, edited by Ruth Mace, Clare J. Holden and Stephen Shennan, 69–85. Tuscon AZ: Left Coast Press.

Buol, Stanley Walter. 2002. "Philosophies of Soil Classifications: From Is to Does." In *Soil Classification: A Global Desk Reference*, edited by Thomas Rice, Hari Eswaran, Bobby A . Stewart and Robert Ahrens, 3–10. Boca Raton, FL: CRC Press.

Buol, Stanley Walter, Randal J. Southard, Robert C. Graham, and Paul A. McDaniel. 2003. *Soil Genesis and Classification.* 5th edn. Ames, Iowa: Iowa State Press.

Cain, A.J. 1995. "Linnaeus's natural and artificial arrangements of plants." *Botanical Journal of the Linnean Society* 117 (2): 73.

Cain, Arthur J. 1999. "Thomas Sydenham, John Ray, and Some Contemporaries on Species." *Archives of Natural History* 24 (1): 55–83.

Cain, Arthur J. 1959. "The Post-Linnaean Development of Taxonomy." *Proceedings of the Linnean Society of London* 170: 234–244.

Candolle, Augustine-Pyramus de. 1819. *Théorie élementaire de la botanique, ou exposition des principes de la classification naturelle et de l'art de décrire et d'étudier les végétaux.* 2nd edn. Paris: Déterville.

Chalmers, A.F. 1990. *Science and its Fabrication.* Milton Keynes: Open University Press.

Charles, David. 2002. *Aristotle on Meaning and Essence.* Oxford: Oxford University Press.

Chevalier, Jacques. 1937. *Claude Bernard: philosophie, manuscrit inédit.* Paris: Boivin.

Chung, Carl. 2003. "On the Origin of the Typological/Population Distinction in Ernst Mayr's Changing Views of Species, 1942–1959." *Studies in History and Philosophy of Biological and Biomedical Sciences* 34: 277–296.

Claridge, Michael F., H.A. Dawah, and M.R. Wilson. 1997. *Species: The Units of Biodiversity.* London: Chapman and Hall.

Coggon, Jennifer. 2002. "Quinarianism after Darwin's Origin: The Circular System of William Hincks." *Journal of the History of Biology* 35 (1): 5.

Collazo, Andres. 2000. "Developmental Variation, Homology, and the Pharyngula Stage." *Systematic Biology* 49 (1): 3–18.

Colless, Donald. 2006. "Taxa, Individuals, Clusters and a few other Things." *Biology and Philosophy* 21 (3): 353–367.

Comte, Auguste. 1853. *The Postive Philosophy of Auguste Comte.* Translated by Harriet Martineau. 2 vols. London: John Chapman. Reprint, Thoemmes Press, 2001.

Cotard, Henri. 1944. *Pour connaître la pensée de Claude Bernard.* Grenoble: Editions Françaises Nouvelles.

Cracraft, Joel. 2005. "Phylogeny and Evo-devo: Characters, Homology, and the Historical Analysis of the Evolution of Development." *Zoology* 108 (4): 345–356.

Cracraft, Joel. 2000. "Species Concepts in Theoretical and Applied Biology: A Systematic Debate with Consequences." In *Species Concepts and Phylogenetic Theory: A Debate,* edited by Quentin D. Wheeler and Rudolf Meier, 3–14. New York: Columbia University Press.

Cracraft, Joel. 1967. "Comments on Homology and Analogy." *Systematic Zoology* 16 (4): 355–359.

Cunningham, Suzanne. 1996. *Philosophy and the Darwinian Legacy.* Rochester: University of Rochester Press.

Darlington, P.J., Jr. 1970. "A Practical Criticism of Hennig-Brundin 'Phylogentic Systematics' and Antarctic Biogeography." *Systematic Zoology* 19 (1): 1–18.

Darwin, Charles Robert. 1859. *On the Origin of Species by Means of Natural Selection, or The Preservation of Favoured Races in the Struggle for Life.* London: John Murray.

Daston, Lorraine, and Katharine Park. 1998. *Wonders and the Order of Nature, 1150–1750.* New York: Zone Books.

de Beer, Gavin Rylands. 1971. *Homology: An Unsolved Problem, Oxford Biology Readers.* London: Oxford University Press.

de Queiroz, Kevin. 1988. "Systematics and the Darwinian Revolution." *Philosophy of Science* 55 (2): 238.

de Queiroz, Kevin, and Jacques Gauthier. 1992. "Phylogenetic Taxonomy." *Annual Review of Ecology and Systematics* 23 (1): 449–480.

de Waal, Frans. 2001. *The Ape and the Sushi Master: Cultural Reflections by a Primatologist.* New York: Basic Books.

de Waal, Frans. 1982. *Chimpanzee Politics: Power and Sex among Apes.* London: Cape.

de Waal, Frans, Stephen Macedo, Josiah Ober, and Christine M. Korsgaard. 2006. *Primates and Philosophers: How Morality Evolved.* Princeton, N.J.: Princeton University Press.

DeSalle, Rob, Mary G. Egan, and Mark Siddall. 2005. "The Unholy Trinity: Taxonomy, Species Delimitation and DNA Barcoding." *Philosophical Transactions: Biological Sciences* 360 (1462): 1905–1916.

Devitt, Michael. 2008. "Resurrecting Biological Essentialism." *Philosophy of Science* 75 (3): 344–382.

Donoghue, Michael J. 1992. "Homology." In *Keywords in Evolutionary Biology,* edited by Evelyn Fox Keller and Elisabeth A. Lloyd, 170–179. Cambridge, MA: Harvard University Press.

Doolittle, W. Ford. 2009. "The Practice of Classification and the Theory of Evolution, and What the Demise of Charles Darwin's Tree of Life Hypothesis means for both of Them." *Philosophical Transactions of the Royal Society B: Biological Sciences* 364 (1527): 2221–2228.

Dott, Robert H. 1998. "What is Unique about Geological Reasoning?" *GSA Today* October 1998: 15–18.

Duhem, Pierre Maurice Marie. 1954[1991]. *The Aim and Structure of Physical Theory*. Princeton: Princeton University Press.

Dunne, Robert A. 2007. *A Statistical Approach to Neural Networks for Pattern Recognition*, Wiley series in computational statistics. Hoboken, NJ: John Wiley and Sons, Inc.

Dupré, John. 2006. "Scientific Classification." *Theory Culture Society* 23 (2–3): 30–32.

Dupré, John. 1993. *The Disorder of Things: Metaphysical Foundations of the Disunity of Science*. Cambridge, MA: Harvard University Press.

Ebach, M. 2005. "Die Anschauung and the Archetype: The Role of Goethe's Delicate Empiricism in Comparative Biology." *Janus Head* 8: 254–270.

Ebach, Malte C., and Craig Holdrege. 2005. "More Taxonomy, Not DNA Barcoding." *Bioscience* 55 (10): 823–824.

Ebach, Malte C., and David M. Williams. 2010. "Aphyly: A Systematic Designation for a Taxonomic Problem." *Evolutionary Biology* 37: 123–127.

Ebach, Malte C., David M. Williams, and Tegan A. Vanderlaan. 2013. "Implementation as Theory, Hierarchy as Transformation, Homology as Synapomorphy." *Zootaxa* 3641: 587.

Ehrlich, Paul R. 1961. "Has the Biological Species Concept Outlived Its Usefulness?" *Systematic Zoology* 10 (4): 167–176.

Elkana, Yehuda. 1984. *William Whewell: Selected writings on the History of Science*. Edited by John Clive, Classics of British historical literature. Chicago: University of Chicago Press.

Ellis, Brian David. 2002. *The Philosophy of Nature: A Guide to the New Essentialism*. Chesham: Acumen.

Ellis, Brian David. 2001. *Scientific Essentialism*, Cambridge studies in philosophy. Cambridge: Cambridge University Press.

Endler, John A. 1986. *Natural Selection in the Wild, Monographs in Population Biology*. Princeton, N.J.: Princeton University Press.

Erickson, Mark. 2012. "Network as Metaphor." *International Journal of Criminology and Sociological Theory* 5 (2): 912–921.

Ereshefsky, Marc. 2010a. "Darwin's Solution to the Species Problem." *Synthese* 175 (3): 405–425.

Ereshefsky, Marc. 2010b. "What's Wrong with the New Biological Essentialism." *Philosophy of Science* 77 (5): 674–685.

Ereshefsky, Marc. 2007. "Psychological Categories as Homologies: Lessons from Ethology." *Biology and Philosophy* 22 (5): 659–674.

Ereshefsky, Marc. 1992. *The Units of Evolution: Essays on the Nature of Species*. Cambridge, MA: MIT Press.

Ereshefsky, Marc. 1991. "Species, Higher Taxa, and the Units of Evolution." *Philosophy of Science* 58: 84–101.

Faith, Daniel P., and Peter S. Cranston. 1992. "Probability, Parsimony, and Popper." *Systematic Biology* 41 (2): 252–257.

Felsenstein, Joseph. 2004. *Inferring Phylogenies*. Sunderland, Mass.: Sinauer Associates.

Fitch, Walter M. 2000. "Homology: A Personal View on Some of the Problems." *Trends in Genetics* 16: 227–231.

Fitzhugh, Kirk. 2009. "Species as Explanatory Hypotheses: Refinements and Implications." *Acta Biotheoretica* 57 (1): 201–248.

Floridi, Luciano. 2004. "Open Problems in the Philosophy of Information." *Metaphilosophy* 35 (4): 554–582.

Floyer, John, and Edward Tyson. 1699. "A Relation of Two Monstrous Pigs, with the Resemblance of Humane Faces, and Two Young Turkeys Joined by the Breast." *Philosophical Transactions of the Royal Society of London* 21: 431–435.

Franz, Nico M. 2005a. "On the Lack of Good Scientific Reasons for the Growing Phylogeny/Classification Gap." *Cladistics* 21 (5): 495–500.

Franz, Nico M. 2005b. "Outline of an Explanatory Account of Cladistic Practice." *Biology & Philosophy* 20 (2–3): 489–515.

Freudenstein, John V. 2005. "Characters, States, and Homology." *Systematic Biology* 54 (6): 965.

Gavrilets, Sergey. 2004. *Fitness Landscapes and the Origin of Species, Monographs in Population Biology*; vol. 41. Princeton, N.J.: Princeton University Press.

Gayon, Jean. 1996. "The Individuality of the Species: A Darwinian Theory? – From Buffon to Ghiselin, and Back to Darwin." *Biology and Philosophy* 11: 215–244.

Gelman, Susan A. 2003. *The Essential Child: Origins of Essentialism in Everyday Thought*, Oxford series in cognitive development. New York: Oxford University Press.

Ghiselin, Michael T. 2005. "Homology as a Relation of Correspondence between Parts of Individuals." *Theory in Biosciences* 124 (2): 91–103.

Ghiselin, Michael T. 1997. *Metaphysics and the Origin of Species*. Albany: State University of New York Press.

Ghiselin, Michael T. 1974. "A Radical Solution to the Species Problem." *Systematic Zoology* 23: 536–544.

Ghiselin, Michael T. 1969. "The Distinction Between Similarity and Homology." *Systematic Zoology* 18 (1): 148–149.

Glen, Christopher L., and Michael B. Bennett. 2007. "Foraging Modes of Mesozoic Birds and Non-avian Theropods." *Current biology CB* 17 (21): R911–R912.

Gliboff, Sander. 2007. "H. G. Bronn and the History of Nature." *Journal of the History of Biology* 40 (2): 259–294.

Godfrey-Smith, Peter. 2003a. "Goodman's Problem and Scientific Methodology." *The Journal of Philosophy* 100 (11): 573–590.

Godfrey-Smith, Peter. 2003b. *Theory and Reality: An Introduction to the Philosophy of Science*. Chicago: University of Chicago Press.

Goethe, Johann Wolfgang von. 1988. *Scientific Studies*. Edited by Douglas Miller. New York, N.Y.: Suhrkamp.

Goethe, Johann Wolfgang von. 1825 (1970). "Versuch einer Witterungslehre." In *Die Schriften zur Naturwissenschaft*, edited by D. Kuhn and W. von Engelhardt, 244–268. Weimer: Hermann Böhlaus Nachfolger.

Good, Ronald. 1935. "The Real Species Problem." *Proceedings of the Linnean Society of London* 147: 107–110.

Goodman, Nelson. 1972. *Problems and Projects*. Indianapolis: Bobbs-Merrill.

Goodman, Nelson. 1954. *Fact, Fiction and Forecast*. London: University of London, The Athlone Press.

Gould, Stephen Jay. 2002. *The Structure of Evolutionary Theory*. Cambridge, MA: The Belknap Press of Harvard University Press.

Gould, Stephen Jay. 1997. "Redrafting the Tree of Life." *Proceedings of the American Philosophical Society* 141 (1): 30–54.

Gould, Stephen Jay. 1996. *Full House: The Spread of Excellence from Plato to Darwin.* New York: Harmony Books.

Gould, Stephen Jay. 1991. *Bully for Brontosaurus: Further Reflections in Natural History.* London: Penguin. Reprint, 1992.

Gould, Stephen Jay. 1988. *Time's Arrow, Time's Cycle: Myth and Metaphor in the Discovery of Geological Time.* London: Penguin. Reprint, 1987.

Gould, Stephen Jay. 1981. *The Mismeasure of Man.* New York: Norton.

Gould, Stephen Jay. 1977. *Ontogeny and Phylogeny.* Cambridge, MA: The Belknap Press of Harvard University Press.

Gray, Asa. 1879. *Structural Botany, or Organography on the Basis of Morphology. To which is added the principles of taxonomy and phytography, and a glossary of botanical terms.* New York, Chicago: Ivison, Blakeman, Taylor.

Gray, G.S, and W.M Fitch. 1983. "Evolution of Antibiotic Resistance Genes: The DNA Sequence of a Kanamycin Resistance Gene from Staphylococcus Aureus." *Molecular Biology and Evolution* 1 (1): 57–66.

Gray, Russell D., and Quentin D. Atkinson. 2003. "Language-tree Divergence Times Support the Anatolian Theory of Indo-European Origin." *Nature* 426 (6965): 435–439.

Green, David M. 2005. "Designatable Units for Status Assessment of Endangered Species." *Conservation Biology* 19 (6): 1813–1820.

Gregg, John Richard. 1954. *The Language of Taxonomy: An Application of Symbolic Logic to the Study of Classificatory Systems.* New York: Columbia University Press.

Griffiths, Paul. 2009. "In What Sense Does 'Nothing Make Sense Except in the Light of Evolution'?" *Acta Biotheoretica* 57 (1): 11–32.

Griffiths, Paul E, and John S. Wilkins. In Press. "When do Evolutionary Explanations of Belief Debunk Belief?" In *Darwin in the 21st Century: Nature, Humanity, and God,* edited by Phillip R. Sloan. Notre Dame, IN: Notre Dame University Press.

Griffiths, Paul E. 2007. "The phenomena of homology." *Biology and Philosophy* 22 (5): 643–658.

Griffiths, Paul E. 2006. "Function, Homology, and Character Individuation." *Philosophy of Science* 73 (1): 1–25.

Griffiths, Paul E. 1999. "Squaring the circle: Natural kinds with Historical Essences." In *Species, New Interdisciplinary Essays,* edited by Richard A. Wilson, 209–228. Cambridge, MA: Bradford/MIT Press.

Griffiths, Paul E. 1996. "Darwinism, Process Structuralism, and Natural Kinds." *Philosophy of Science* 63 (3 Suppl S): 0031–8248.

Hackett, Shannon J., Rebecca T. Kimball, Sushma Reddy, Rauri C. K. Bowie, Edward L. Braun, Michael J. Braun, Jena L. Chojnowski, W. Andrew Cox, Kin-Lan Han, John Harshman, Christopher J. Huddleston, Ben D. Marks, Kathleen J. Miglia, William S. Moore, Frederick H. Sheldon, David W. Steadman, Christopher C. Witt, and Tamaki Yuri. 2008. "A Phylogenomic Study of Birds Reveals Their Evolutionary History." *Science* 320 (5884): 1763–1768.

Hacking, Ian. 2007. "Natural Kinds: Rosy Dawn, Scholastic Twilight." *Royal Institute of Philosophy Supplement* 82 (Supplement 61): 203–239.

Hacking, Ian. 1991. "A Tradition of Natural Kinds." *Philosophical Studies* 61: 109–126.

Hacking, Ian. 1990. "Natural Kinds." In *Perspectives on Quine,* edited by Robert B. Barrett and Roger F. Gibson. Cambridge: Blackwell.

Hacking, Ian. 1983. *Representing and Intervening: Introductory Topics in the Philosophy of Natural Science.* Cambridge UK: Cambridge University Press.

Hall, Brian K. 2012. "Homology, Homoplasy, Novelty, and Behavior." *Developmental Psychobiology*

Hall, Brian K. 1999. *Homology, Novartis Foundation Symposium 222.* New York: John Wiley and Sons, Inc.

Hall, Brian K. 1994. *Homology: The Hierarchical Basis of Comparative Biology.* San Diego: Academic Press.

Hamblyn, Richard. 2001. *The Invention of Clouds: How an Amateur Meteorologist Forged the Language of the Skies.* London: Picador.

Hamilton, Terrell H. 1967. *Process and Pattern in Evolution.* New York: Macmillan.

Hamilton, William, Henry Longueville Mansel, and John Veitch. 1874. *Lectures on Metaphysics and Logic.* Edinburgh: Blackwood.

Hanson, N.R. 1958. *Patterns of Discovery.* New York: Cambridge University Press.

Haraway, Donna. 1992. "The Promises of Monsters: A Regenerative Politics for Inappropriate/d Others." In *Cultural Studies,* edited by Lawrence Grossberg, Cary Nelson and Paula A. Treichler, 295–337. New York: Routledge.

Hennig, W. 1975. "'Cladistic Analysis or Cladistic Classification?': A Reply to Ernst Mayr." *Systematic Zoology* 244–256.

Hennig, W. 1966. *Phylogenetic systematics.* Translated by D. Dwight Davis and Rainer Zangerl. Urbana: University of Illinois Press.

Hennig, W. 1965. "Phylogenetic Systematics." *Annual Review of Entomology* 10 (1): 97–116.

Hennig, W. 1950. *Grundzüge einer Theorie der Phylogenetischen Systematik.* Berlin: Aufbau Verlag.

Henrich, Joseph, Steven J. Heine, and Ara Norenzayan. 2010. "The Weirdest People in the World?" *Behavioral and Brain Sciences* 33 (2–3): 61–83.

Herdman, William Abbott. 1885. *A Phylogenetic Classification of Animals (For the Use of Students).* London: Macmillan.

Hillis, David M. 1994. "Homology in Molecular Biology." In *Homology, the Hierarchical Basis of Comparative Biology,* edited by Brian K. Hall, 339–368. San Diego: Academic Press.

Hillis, David M., Craig Moritz, Barbara K. Mable, and Axel Meyer. 1996. "Molecular Systematics (2nd edition)." *Trends in Genetics* 12 (12): 534–534.

Hoenigswald, Henry M. 1987. "Language Family Trees: Topological and Metrical." In *Biological Metaphor and Cladistic Classification,* edited by Henry M. Hoenigswald and Linda F. Wiener, 39–80. Philadelphia PA: University of Pennsylvania Press.

Hoßfeld, Uwe, and Lennart Olsson. 2005. "The History of the Homology Concept and the 'Phylogenetisches Symposium'." *Theory in Biosciences* 124 (2): 243–253.

Hull, David L. 1999. "On the Plurality of Species: Questioning the Party Line." In *Species, New Interdisciplinary Essays,* edited by R.A. Wilson, 23–48. Cambridge, MA: Bradford/MIT Press.

Hull, David L. 1994. "Ernst Mayr, Influence on the History and Philosophy of Biology – a Personal Memoir." *Biology and Philosophy* 9 (3): 375–386.

Hull, David L. 1992. "Individual." In *Keywords in Evolutionary Biology*, edited by E.F. Keller and E.A. Lloyd, 180–187. Cambridge MA: Harvard University Press.

Hull, David L. 1988. *Science as a Process: An Evolutionary Account of the Social and Conceptual Development of Science*. Chicago: University of Chicago Press.

Hull, David L. 1984. "Historical Entities and Historical Narratives." In *Minds, Machines, and Evolution*, edited by C. Hookway. Cambridge: Cambridge University Press.

Hull, David L. 1981. "Units of Evolution: A Metaphysical Essay." In *The Philosophy of Evolution*, edited by U.L. Jensen and R. Harré, 23–44. Brighton UK: Harvester Press.

Hull, David L. 1978. "A Matter of Individuality." *Philosophy of Science* 45: 335–360.

Hull, David L. 1965. "The Effect of Essentialism on Taxonomy: Two Thousand Years of Stasis." *British Journal for the Philosophy of Science* 15: 314–326, 16: 1–18.

Hull, David L., and John S. Wilkins. 2005. "Replication." *Stanford Encyclopedia of Philosophy*, http://plato.stanford.edu/entries/replication/.

Hume, David. 1779. *Dialogues Concerning Natural Religion*. 2nd edn. London: Unknown.

Huxley, Julian. 1940. *The New Systematics*. London: Oxford University Press.

Inglis, William G. 1970. "Similarity and Homology." *Systematic Zoology* 19 (1): 93.

Inglis, William G. 1966. "The Observational Basis of Homology." *Systematic Zoology* 15 (3): 219–228.

Jaramillo, M. Alejandra, and Elena M. Kramer. 2007. "The Role of Developmental Genetics in Understanding Homology and Morphological Evolution in Plants." *International Journal of Plant Sciences* 168 (1): 61–72.

Jardine, N. 1967. "The Concept of Homology in Biology." *The British Journal for the Philosophy of Science* 18 (2): 125–139.

Jevons, W. Stanley. 1878. *The Principles of Science: A Treatise on Logic and Scientific Method*. 2nd edn. London: Macmillan, Original edition, 1873.

Jevons, W. Stanley. 1958. *The Principles of Science: A Treatise on Logic and Scientific Method*. Dover, NY.

Johnson, Kristin. 2007. "Natural History as Stamp Collecting: A Brief History." *Archives of Natural History* 34 (2): 244–258.

Junker, T. 1991. "Heinrich Georg Bronn und Origin of Species." *Sudhoffs Archiv; Zeitschrift für Wissenschaftsgeschichte. Beihefte* 75 (2): 180–208.

Katz, S. 1983. "R L Gregory and Others: The Wrong Picture of the Picture Theory of Perception." *Perception* 12 (3): 269–279.

Kauffman, Stuart A. 1993. *The Origins of Order: Self-organization and Selection in Evolution*. New York: Oxford University Press.

Keller, Roberto A., Richard N. Boyd, and Quentin D. Wheeler. 2003. "The Illogical Basis of Phylogenetic Nomenclature." *The Botanical Review* 69 (1): 93–110.

Key, K.H.L. 1967. "Operational Homology." *Systematic Zoology* 16 (3): 275–276.

Kitcher, Philip. 1989. "Some Puzzles about Species." In *What the Philosophy of Biology is: Essays Dedicated to David Hull*, edited by M. Ruse, 183–208. Dordrecht: Kluwer.

Kitching, Ian J., Peter L. Forey, Christopher J. Humphries, and David M. Williams. 1998. *Cladistics: The Theory and Practice of Parsimony Analysis*. 2nd edn. Vol. 11, Systematics Association Publications. New York: Oxford University Press.

Kitts, David B, and David J Kitts. 1979. "Biological Species as Natural Kinds." *Philosophy of Science* 46: 613–622.

Kluge, Arnold. 2009. "Explanation and Falsification in Phylogenetic Inference: Exercises in Popperian Philosophy." *Acta Biotheoretica* 57 (1): 171–186.

Knowles, L. Lacey, and Bryan C. Carstens. 2007. "Delimiting Species without Monophyletic Gene Trees." *Systematic Biology* 56 (6): 887–895.

Kress, W. John, Kenneth J. Wurdack, Elizabeth A. Zimmer, Lee A. Weigt, and Daniel H. Janzen. 2005. "Use of DNA Barcodes to Identify Flowering Plants." *PNAS* 102 (23): 8369–8374.

Kuhn, Thomas S. 1996. *The Structure of Scientific Revolutions*. 3rd edn. Chicago, IL: University of Chicago Press.

Kuhn, Thomas S. 1977. *The Essential Tension: Selected Studies in Scientific Tradition and Change*. Chicago: University of Chicago Press.

Kuhn, Thomas S. 1970. *The Structure of Scientific Revolutions. 2nd enl. ed, International Encyclopedia of Unified Science. Foundations of the Unity of Science*; Vol. 2, No. 2. Chicago: University of Chicago Press.

Kuhn, Thomas S. 1962. *The Structure of Scientific Revolutions, International Encyclopedia of Unified Science*; Vol. 2, No. 2. Chicago: University of Chicago Press.

Kultgen, J.H. 1958. "Philosophic Conceptions in Mendeleev's Principles of Chemistry." *Philosophy of Science* 25 (3): 177–183.

Kuntz, Marion Leathers, and Paul Grimley Kuntz. 1988. *Jacob's Ladder and the Tree of Life: Concepts of Hierarchy and the Great Chain of Being*. Revised Edn. Vol. 14, American University Studies. Series V, Philosophy. New York: P. Lang.

Kupfer, David J., and Darrel A. Regier. 2011. "Neuroscience, Clinical Evidence, and the Future of Psychiatric Classification in DSM-5." *American Journal of Psychiatry* 168 (7): 672–674.

Kurt Lienau, E., and Rob DeSalle. 2009. "Evidence, Content and Corroboration and the Tree of Life." *Acta Biotheoretica* 57 (1): 187–199.

Kutzbach, Gisela. 1979. *The Thermal Theory of Cyclones: A History of Meteorological Thought in the Nineteenth Century*. Boston: American Meteorological Society.

Laet, Jan, and Erik Smets. 1998. "On the Three-Taxon Approach to Parsimony Analysis." *Cladistics* 14 (4): 363–381.

Lakatos, Imre. 1976. *Proofs and Refutations: The Logic of Mathematical Discovery*. Cambridge; New York: Cambridge University Press.

Lakatos, Imre. 1970. "Falsification and the Methodology of Scientific Research Programmes." In *Criticism and the Growth of Knowledge*, edited by Imre Lakatos and Alan Musgrave, 91–196. London: Cambridge University Press.

Lankester, Edwin Ray. 1870. "On the Use of the Term Homology in Modern Zoology, and the Distinction between Homogenetic and Homoplastic Agreements." *Annals and Magazine of Natural History* 4 (6): 34–43.

LaPorte, Joseph. 2004. *Natural Kinds and Conceptual Change*. Cambridge UK: Cambridge University Press.

LaPorte, Joe. 2003. "Does a Type Specimen Necessarily or Contingently belong to its Species?" *Biology and Philosophy* 18: 583–588.

LaPorte, Joseph. 1997. "Essential Membership." *Philosophy of Science* 64: 96–112.

LaPorte, Joe. 1996. "Chemical Kind Term Reference and the Discovery of Essence." *Nous*

Laubichler, Manfred D. 2000. "Homology in Development and the Development of the Homology Concept." *AAmerican Zoologist* 40 (5): 777–788.

Lawless, Kimberly A., and Jonna M. Kulikowich. 2006. "Domain Knowledge and Individual Interest: The Effects of Academic Level and Specialization in Statistics and Psychology." *Contemporary Educational Psychology* 31 (1): 30–43.

Leibniz, Gottfried Wilhelm. 1996. *New Essays on Human Understanding*. Translated by Peter Remnant and Jonathon Bennett. Cambridge UK: Cambridge University Press. Original edition, 1765.

Lennox, James G. 2002. *Aristotle: On the Parts of Animals I-IV*. New York: Oxford University Press, USA.

Lennox, James G. 2001. *Aristotle's Philosophy of Biology: Studies in the Origins of Life Science*. Cambridge, UK: Cambridge University Press.

Levine, Alex. 2001. "Individualism, Type Specimens, and the Scrutability of Species Membership." *Biology and Philosophy* 16: 325–338.

Levit, Georgy S., and Kay Meister. 2006. "The History of Essentialism vs. Ernst Mayr's 'Essentialism Story': A Case Study of German Idealistic Morphology" *Theory in Biosciences* 124: 281–307.

Lewis, David K. 1970. "How to Define Theoretical Terms." *The Journal of Philosophy* 67 (13): 427–446.

Lewis, David K. 1991. *Parts of Classes*. Oxford UK: Basil Blackwell.

Lewis, David K. 1969. *Convention: A Philosophical Study*. Cambridge: Harvard University Press.

Lindley, John. 1830. *An Introduction to the Natural System of Botany: Or, a Systematic View of the Organisation, Natural Affinities, and Geographical Distribution, of the Whole Vegetable Kingdom: Together with the Uses of the Most Important Species in Medicine, the Arts, and Rural or Domestic Economy*. 1st edn. London: Longman, Rees, Orme, Brown, and Green.

Lindroth, Sten. 1983. "The Two Faces of Linnaeus." In *Linnaeus, the Man and his Work*, edited by Tore Frängsmyr, 1–62. Berkeley: University of California Press.

Linnaeus, Carl. 1758–1759. *Systema naturae per regna tria naturae: secundum classes, ordines, genera, species, cum characteribus, differentiis, synonymis, locis*. Holmiae: Laurentii Salvii.

Lipton, Peter. 1991. "Contrastive Explanation and Causal Triangulation." *Philosophy of Science* 58 (4): 687–697.

Lipton, Peter. 1990. "Contrastive Explanation." *Royal Institute of Philosophy Supplements* 27 (1): 247–266.

Locke, John. c1900. *An Essay Concerning Human Understanding*. 6th edn. London: Ward Lock. Original edition, 1690.

Lorenz, Konrad, and Agnes von Cranach. 1996. *The Natural Science of the Human Species: An Introduction to Comparative Behavioral Research: The "Russian Manuscript" (1944–1948)*. Cambridge, MA: MIT Press.

Losos, Jonathan B., David M. Hillis, and Harry W. Greene. 2012. "Who Speaks with a Forked Tongue?" *Science* 338 (6113): 1428–1429.

Love, Alan. 2007. "Functional Homology and Homology of Function: Biological Concepts and Philosophical Consequences." *Biology and Philosophy* 22 (5): 691–708.

Lovejoy, Arthur O. 1936. *The Great Chain of Being: A Study of the History of an Idea*. Cambridge, MA.: Harvard University Press. Reprint, 1964.

Lovejoy, Arthur O. 1904. "Some Eighteenth Century Evolutionists. II." *Popular Science Monthly* LXV (August): 323–340.

Mackintosh, Robert. 1899. *From Comte to Benjamin Kidd: The Appeal to Biology or Evolution for Human Guidance.* London: Macmillan & Co., Ltd.

Maclaurin, James, and Kim Sterelny. 2008. *What is Biodiversity?* Chicago: University of Chicago Press.

Macleay, William Sharp. 1819. *Horae entomologicae, or, Essays on the annulose animals.* London: Printed for S. Bagster.

Magnus, P.D. 2012. *Scientific Enquiry and Natural Kinds: From Planets to Mallards.* Basingstoke: Palgrave Macmillan.

Mahoney, Anne. 2008. Review of R.S.P. Beekes, "The Origin of the Etruscans." *Etruscan Studies* 11: Article 12, http://scholarworks.umass.edu/etruscan_studies/vol11/iss1/12.

Makarenkov, V., and P. Legendre. 2004. "From a Phylogenetic Tree to a Reticulated Network." *Journal of Computational Biology* 11 (1): 195–212.

Marbut, Curtis F. 1922. "Soil Classification." Bulletin of the American Association of Soil Survey Workers Bull 3

Marcuse, Herbert. 1964. *One-Dimensional Man: Studies in the Ideology of Advanced Industrial Society.* London: Routledge and Kegan Paul.

Massimi, Michela. 2011. "From Data to Phenomena: A Kantian Stance." *Synthese* 182 (1): 101–116.

Massimi, Michela. 2008a. *Kant and Philosophy of Science Today, Royal Institute of Philosophy Supplements: 63.* Cambridge: Cambridge University Press.

Massimi, Michela. 2008b. "Why There are no Ready-made Phenomena: What Philosophers of Science Should Learn from Kant." *Kant and Philosophy of Science Today, Royal Institute of Philosophy Supplement* 63:1–35.

Masterson, Margaret. 1970. "The Nature of a Paradigm." In *Criticism and the Growth of Knowledge*, edited by Imre Lakatos and Alan Musgrave, 59–89. Cambridge: Cambridge University Press.

Mauriac, Pierre. 1940. *Claude Bernard.* Paris: B. Grasset.

Mayden, R.L., and R.M. Wood. 1995. "Systematics, Species Concepts, and the Evolutionarily Significant Unit in Biodiversity and Conservation Biology." In *American Fisheries Society Symposium* 1995, edited by J.L. Nielsen.

Mayden, Richard L. 1992. *Systematics, Historical Ecology, and North American Freshwater Fishes.* Stanford, Calif.: Stanford University Press.

Mayer, Klaus, Maria Wallenius, and Ian Ray. 2005. "Nuclear Forensics-a Methodology Providing Clues on the Origin of Illicitly Trafficked Nuclear Materials." *Analyst* 130 (4): 433–441.

Mayhew, Robert. 2004. *The Female in Aristotle's Biology: Reason or Rationalization.* Chicago; London: University of Chicago Press.

Maynard Smith, John. 1958. *The Theory of Evolution.* Harmondsworth, UK: Penguin Books.

Mayr, Ernst. 1982. *The Growth of Biological Thought: Diversity, Evolution, and Inheritance.* Cambridge, MA: The Belknap Press of Harvard University Press.

Mayr, Ernst. 1974. "Cladistic Analysis or Cladistic Classification?" *Journal of Zoological Systematics and Evolutionary Research* 12 (1): 94–128.

Mayr, Ernst. 1969. "The Biological Meaning of Species." *Biological Journal of the Linnean Society* 1 (3): 311–320.

Mayr, Ernst. 1949. "Speciation and Selection." *Proceedings of the American Philosophical Society* 93 (6): 514–519.

McDonald, Robert C., R.F. Isbell, James G. Speight, and J. Walker. 1984. *Australian Soil and Land Survey: Field Handbook.* Melbourne: Inkata Press.

McHenry, Leemon B. 2000. "Review: The Compressibility of the Universe: A New Conception of Science, by Nicholas Maxwell." *Mind* 109 (433): 162–166.

McKitrick, Mary C. 1994. "On Homology and the Ontological Relationship of Parts." *Systematic Biology* 43 (1): 1–10.

McLaughlin, Peter. 2002. "Naming Biology." *Journal of the History of Biology* 35 (1): 1–4.

McOuat, Gordon. 2009. "The Origins of 'Natural Kinds': Keeping 'Essentialism' at Bay in the Age of Reform." *Intellectual History Review* 19 (2): 211–230.

McOuat, Gordon R. 2003. "The Logical Systematist: George Bentham and his Outline of a New System of Logic." *Archives of Natural History* 30 (2): 203–223.

Medawar, Peter Brian. 1969. *Induction and Intuition in Scientific Thought.* London: Methuen.

Mill, John Stuart. 1974. *A System of Logic, Ratiocinative and Inductive: Being a Connected View of the Principles of Evidence and the Methods of Scientific Investigation,* Books I–III. Edited by J.M. Robson. Vol. VII, *Collected Works of John Stuart Mill.* Toronto, Buffalo NY, London: University of Toronto Press/Routledge and Kegan Paul.

Mindell, David P., and Axel Meyer. 2001. "Homology Evolving." *Trends in Ecology and Evolution* 16 (8): 434–440.

Mittelstrass, Jürgen, Peter McLaughlin, and Burgen, Arnold S.V. 1997. "The Idea of Progress." *Philosophie und Wissenschaft, transdisziplinäre Studien;* Bd. 13. New York: W. de Gruyter.

Mommsen, Theodor. 1862. *The History of Rome.* Translated by William Purdie Dickson and Leonhard Schmitz. Vol. 1. London: R. Bentley.

Monk, Ray. 1990. *Ludwig Wittgenstein: The Duty of Genius.* 1st American ed. New York: Free Press.

Moreno, Luis Fernandez. 1999. "Reference Change of Natural Kind Terms." *Sorites*

Mößner, Nicola. 2011. "Thought Styles and Paradigms – A Comparative Study of Ludwik Fleck and Thomas S. Kuhn." *Studies In History and Philosophy of Science Part A* 42 (2): 362–371.

Mückenhausen, E. 1965. "The Soil Classification System of the Federal Republic of Germany." *Pédologie,* Gand, numéro spécial 3: 57–89

Mufwene, Salikoko S. 2001. *The Ecology of Language Evolution, Cambridge Approaches to Language Contact.* Cambridge, UK: Cambridge University Press.

Müller, Gerd B. 2003. "Homology: The Evolution of Morphological Organization." In *Origination of Organismal Form: Beyond the Gene in Developmental and Evolutionary Biology,* edited by Gerd B. Müller and Stuart A. Newman, 51–70. Cambridge, MA: MIT Press.

Muller, Gerd B., and Gunter P. Wagner. 1996. "Homology, Hox Genes, and Developmental Integration." *American Zoologist* 36 (1): 4–13.

Müller-Wille, Staffan. 2007. "Collection and Collation: Theory and Practice of Linnaean Botany." *Studies in History and Philosophy of Science Part C: Studies in History and Philosophy of Biological and Biomedical Sciences* 38 (3): 541–562.

Müller-Wille, Staffan. 2003. "Nature as a Marketplace: The Political Economy of Linnaean Botany." *History of Political Economy* 35 (Annual Supplement): 154–172.

Murphy, Dominic. 2006. *Psychiatry in the Scientific Image.* Cambridge, MA: MIT Press.

Nelson, Gareth J. 2011. "Resemblance as Evidence of Ancestry." *Zootaxa* 2946: 137–141.

Nelson, Gareth J. 1985. "Class and Individual: A Reply to M. Ghiselin." *Cladistics-The International Journal of the Willi Hennig Society* 1: 386–389.

Nelson, Gareth J. 1978. "Professor Michener on Phenetics – Old and New." *Systematic Zoology* 27 (1): 104–112.

Nelson, Gareth J. 1973. "Classification as an Expression of Phylogenetic Relationships." *Systematic Zoology* 22 (4): 344–359.

Nelson, Gareth J., and Pauline Y. Ladiges. 2009. "Biogeography and the Molecular Dating Game: A Futile Revival of Phenetics?" Bulletin de la Société Géologique de France 180 (1): 39–43.

Nelson, Gareth J., and Norman I. Platnick. 1981. *Systematics and Biogeography: Cladistics and Vicariance.* New York: Columbia University Press.

Nietzsche, Friedrich Wilhelm. 1966. *Beyond Good and Evil: Prelude to a Philosophy of the Future.* Translated by Walter Kaufmann. New York, NY: Random.

Nixon, Kevin C., and James M. Carpenter. 2011. "On Homology." *Cladistics: Early View.*

O'Hara, Robert J. 1997. "Population Thinking and Tree Thinking in Systematics." *Zoologica Scripta* 26 (4): 323–329.

O'Hara, Robert J. 1996. "Trees of History in Systematics and Philology." *Memorie della Società Italiana di Scienze Naturali e del Museo Civico di Storia Naturale di Milano* 27 (1): 81–88.

O'Hara, Robert J. 1993. "Systematic Generalization, Historical Fate, and the Species Problem." *Systematic Biology* 42 (3): 231–246.

O'Hara, Robert J. 1991. "Representations of the Natural System in the Nineteenth Century." *Biology and Philosophy* 6 (2): 255–274.

O'Malley, Maureen A., and Yan Boucher. 2005. "Paradigm Change in Evolutionary Microbiology." *Studies in History and Philosophy of Biological and Biomedical Sciences* 36 (1):183–208.

O'Malley, Maureen A., and John Dupré. 2007. "Size Doesn't Matter: Towards a More Inclusive Philosophy of Biology." *Biology and Philosophy* 22 (2): 155–191.

O'Malley, Maureen A. 2010. "Ernst Mayr, the Tree of Life, and Philosophy of Biology." *Biology and Philosophy* 25 (4): 529–552.

Ochman, Howard, Jeffrey G. Lawrence, and Eduardo A. Groisman. 2000. "Lateral Gene Transfer and the Nature of Bacterial Innovation." *Nature* 405: 299–304.

Ochman, Howard, Emmanuelle Lerat, and Vincent Daubin. 2005. "Examining Bacterial Species Under the Specter of Gene Transfer and Exchange." *PNAS* 102 (Suppl 1): 6595–6599.

Okada, Takeshi, and Herbert A. Simon. 1997. "Collaborative Discovery in a Scientific Domain." *Cognitive Science* 21 (2): 109–146.

Okasha, Samir. 2002. "Darwinian Metaphysics: Species and the Question of Essentialism." *Synthese: An International Journal for Epistemology, Methodology and Philosophy of Science* 131 (2): 191–213.

Opitz, J.M. 2004. "Goethe's Bone and the Beginnings of Morphology." *American Journal of Medical Genetics Part A* 126A (1):1–8.

Ornduff, Robert. 1969. *The Systematics of Populations in Plants*, Systematic Biology Pubn 1692. Washington DC: NAS.

Ostrovsky, Yuri, Aaron Andalman, and Pawan Sinha. 2006. "Vision Following Extended Congenital Blindness." *Psychological Science* 17 (12): 1009–1014.

Owen, Richard. 1859. *On the Classification and Geographical Distribution of the Mammalia*. London: John W. Parker and Son.

Owen, Richard. 1848. *The Archetype and Homologies of the Vertebrate Skeleton*. London: J. van Voorst.

Owen, Richard. 1843. *Lectures on the Comparative Anatomy and Physiology of the Invertebrate Animals*, delivered at the Royal College of Surgeons, in 1843. By Richard Owen. From notes taken by William White Cooper and revised by Professor Owen. London: Longman, Brown, Green, and Longmans.

Page, Timothy J., Satish C. Choy, and Jane M. Hughes. 2005. "The Taxonomic Feedback Loop: Symbiosis of Morphology and Molecules." *Biology Letters* 1 (2): 139–142.

Pallas, Peter Simon. 1766. *Elenchus zoophytorum sistens generum adumbrationes generaliores et specierum cognitarum succintas descriptiones, cum selectis auctorum synonymis*. Hagae-Comitum: Apud Petrum van Cleef.

Panchen, Alec L. 1992. *Classification, Evolution, and the Nature of Biology*. Cambridge UK: Cambridge University Press.

Papineau, David. 1979. *Theory and Meaning*. New York: Oxford University Press.

Park, Katharine, and Lorraine J. Daston. 1981. "Unnatural Conceptions: The Study of Monsters in Sixteenth- and Seventeenth-Century France and England." *Past & Present* 92 (August, 1981): 20–54.

Patterson, Colin. 1999. *Evolution*. 2nd edn. Ithaca, N.Y.: Comstock Pub. Associates.

Patterson, Colin. 1988a. "Homology in Classical and Molecular Biology." *Molecular Biology and Evolution* 5 (6): 603–625.

Patterson, Colin. 1988b. "The Impact of Evolutionary Theories on Systematics." In *Prospects in Systematics*, edited by D.L. Hawksworth. Oxford: Oxford University Press.

Patterson, Colin 1982a. "Morphological Characters and Homology." In *Problems of Phylogenetic Reconstruction*, edited by K.A. Joysey and A.E. Friday, 21–74. London: Academic Press.

Patterson, Colin. 1982b. "Classes and Cladists or Individuals and Evolution." *Systematic Zoology* 31: 284–286.

Paul, D.N. Hebert, and T. Ryan Gregory. 2005. "The Promise of DNA Barcoding for Taxonomy." *Systematic Biology* 54 (5): 852.

Pavord, Anna. 2005. *The Naming of Names: The Search for Order in the World of Plants*. London: Bloomsbury.

Pavord, Anna. 2009. *Searching for Order: The History of the Alchemists, Herbalists and Philosophers who Unlocked the Secrets of the Plant World*. London: Bloomsbury.

Pearson, Christopher H. 2010. "Pattern Cladism, Homology, and Theory-neutrality." *History and Philosophy of the Life Sciences* 32 (4): 475.

Pellecchia, Marco, Riccardo Negrini, Licia Colli, Massimiliano Patrini, Elisabetta Milanesi, Alessandro Achilli, Giorgio Bertorelle, Luigi L. Cavalli-Sforza, Alberto Piazza, Antonio Torroni, and Paolo Ajmone-Marsan. 2007. "The Mystery of

Etruscan Origins: Novel Clues from Bos taurus Mitochondrial DNA." *Proceedings: Biological Sciences* 274 (1614): 1175–1179.

Pellegrin, Pierre. 1986. *Aristotle's Classification of Animals: Biology and the Conceptual Unity of the Aristotelian Corpus*. Translated by Anthony Preus. Revised edn. Berkeley: University of California Press.

Platnick, Norman I. 2012. "Less on Homology." *Cladistics*.

Playfair, John. 1802. *Illustrations of the Huttonian theory*. London: Adell and Davies.

Poincaré, Henri. 2003. *Science and Method*. Dover edn. Mineola, N.Y.: Dover Publications.

Prinz, Jesse J. 2002. *Furnishing the Mind: Concepts and their Perceptual Basis*. Edited by Hilary Putnam and Ned Block, Representation and Mind. Cambridge, MA: MIT Press.

Psillos, Stathis. 1999. *Scientific Realism: How Science Tracks Truth*. New York: Routledge.

Putnam, Hilary. 1981. *Reason, Truth and History*. Cambridge, UK: Cambridge University Press.

Putnam, Hilary. 1975. *Mind, Language, and Reality, Philosophical Papers* v. 2. Cambridge, UK: Cambridge University Press.

Quine, Willard Van Orman. 1993. "In Praise of Observation Sentences." *The Journal of Philosophy* 90 (3): 107–116.

Quine, Willard Van Orman. 1969a. "Natural Kinds." In *Essays in Honour of Carl G. Hempel: A Tribute on the Occasion of His Sixty-Fifth Birthday*, edited by Nicholas Rescher, 5–27. Dordrecht, Holland: Springer.

Quine, Willard Van Orman. 1969b. *Ontological Relativity and Other Essays*. New York: Columbia University Press.

Quine, Willard Van Orman. 1953. *From a Logical Point of View: 9 Logico-philosophical Essays*. Cambridge, MA: Harvard University Press.

Quine, Willard Van Orman. 1948. "On What There Is." *Review of Metaphysics* 2 (5): 21–38.

Ragan, Mark A. 2001. "Detection of Lateral Gene Transfer among Microbial Genomes." *Current Opinion in Genetics and Development* 11 (6): 620–626.

Ragan, Mark A., James O. McInerney, and James A. Lake. 2009. "The Network of Life: Genome Beginnings and Evolution." *Philosophical Transactions of the Royal Society B: Biological Sciences* 364 (1527): 2169–2175.

Ramsey, Frank Plumpton. 1931 (1954). *The Foundations of Mathematics and other Logical Essays*. London: Routledge and Kegan Paul.

Ramsey, Grant, and Anne Siebels Peterson. 2012. "Sameness in Biology." *Philosophy of Science* 79 (2): 255–275.

Redfearn, Suz. 2002. "Conflict Diamonds." *Opt. Photon. News* 13 (2): 20–22.

Regier, Darrel A., William E. Narrow, Emily A. Kuhl, and David J. Kupfer. 2009. "The Conceptual Development of DSM-V." *American Journal of Psychiatry* 166 (6): 645–650.

Reichenbach, Hans. 1949. *The Theory of Probability, an Inquiry into the Logical and Mathematical Foundations of the Calculus of Probability*. 2nd edn. Berkeley: University of California Press.

Rendall, Drew, and Anthony Di Fiore. 2007. "Homoplasy, Homology, and the Perceived Special Status of Behavior in Evolution." *Journal of Human Evolution* 52 (5): 504–521.

Richardson, Ernest Cushing. 1901. *Classification, Theoretical and Practical I. The Order of the Sciences.* 2 vols. Vol. 1. New York: Scribner.

Ridley, Mark. 1986. *Evolution and Classification: The Reformation of Cladism.* London; New York: Longman.

Riedman, Marianne. 1991. *The Pinnipeds: Seals, Sea Lions, and Walruses.* Berkeley: University of California Press.

Rieppel, Olivier. 2009. "Species as a Process." *Acta Biotheoretica* 57 (1):33–49.

Rieppel, Olivier. 2007. "6 Homology: A Philosophical and Biological Perspective." In, 217–240.

Rieppel, Olivier. 2005. "Modules, Kinds, and Homology." *Journal of Experimental Zoology Part B: Molecular and Developmental Evolution* 304B (1): 18–27.

Rieppel, Olivier. 2003a. "Popper and Systematics." *Systematic Biology* 52 (2): 259–271.

Rieppel, Olivier. 2003b. "Semaphoronts, Cladograms and the Roots of Total Evidence." *Biological Journal of the Linnean Society* 80 (1): 167–186.

Rieppel, Olivier. 1994. "Homology, Topology, and Typology: The History of Modern Debates." In *Homology: The Hierarchical Basis of Comparative Biology*, edited by Brian K. Hall, 63–100. San Diego: Academic Press.

Rogers, Steven K., and Matthew Kabrisky. 1991. *An Introduction to Biological and Artificial Neural Networks for Pattern Recognition, Tutorial texts in Optical Engineering.* Bellingham, Wash., USA: SPIE Optical Engineering Press.

Rosenberg, Alexander. 1994. *Instrumental Biology, or, The Disunity of Science.* Chicago: University of Chicago Press.

Rosenberg, Alexander. 2006. *Darwinian Reductionism, or, How to Stop Worrying and Love Molecular Biology.* Chicago: University of Chicago Press.

Rossi, Paolo. 2000. *Logic and the Art of Memory: The Quest for a Universal Language.* London: Athlone.

Ruse, Michael. 1998. "All My Love is Toward Individuals." *Evolution* 52: 283–288.

Ruse, Michael. 1996. *Monad to Man: The Concept of Progress in Evolutionary Biology.* Cambridge, MA.: Harvard University Press.

Ruse, Michael. 1987. "Biological Species: Natural Kinds, Individuals, or What?" *British Journal for the Philosophy of Science* 38: 225–242.

Ruse, Michael. 1969. "Definitions of Species in Biology." *British Journal for the Philosophy of Science* 20: 97–119.

Russell, Bertrand. 1950. *Unpopular Essays.* London: George Allen and Unwin.

Russell, Bertrand. 1919. *Introduction to Mathematical Philosophy.* London; New York: G. Allen and Unwin; MacMillan and Co.

Salmon, Wesley C. 1991. "Hans Reichenbach's Vindication of Induction." *Erkenntnis* 35 (1): 99–122.

Sandri, Giorgio. 1969. "On the Logic of Classification." *Quality and Quantity* 3 (1–2): 80–124.

Sandvik, Hanno. 2009. "Anthropocentricisms in Cladograms." *Biology and Philosophy* 24 (4): 425–440.

Sankey, Howard. 1998. "Taxonomic Incommensurability." *International Studies in the Philosophy of Science* 12 (1): 7–16.

Sankey, Howard. 1994. *The Incommensurability Thesis.* Aldershot; Sydney: Avebury.

Sarkar, Sahotra. 1996. *The Emergence of Logical Empiricism: From 1900 to the Vienna Circle.* New York: Garland Publishing.

Scerri, Eric R. 2012. "A Critique of Weisberg's View on the Periodic Table and Some Speculations on the Nature of Classifications." Foundations of Chemistry

Scerri, Eric R. 2007. *The Periodic Table: Its Story and Its Significance*. New York: Oxford University Press. Original edition, 2006.

Scheffler, Israel. 1967. *Science and Subjectivity*. Indianapolis,: Bobbs-Merrill.

Schlichting, Carl D., and Massimo Pigliucci. 1998. *Phenotypic Evolution: A Reaction Norm Perspective*. Sunderland, MA: Sinauer Associates.

Schuh, Randall T., and Andrew V.Z. Brower. 2009. *Biological Systematics: Principles and Application*. 2nd edn. Ithaca NY: Cornell University Press.

Scotland, Robert W. 2000. "Taxic Homology and Three-Taxon Statement Analysis." *Systematic Biology* 49 (3): 480–500.

Seamon and A. Zajonc. 1998. *Goethe's Way of Science: A Phenomenology of Nature*. New York: SUNY Press.

Sedley, David N. 2007. *Creationism and its Critics in Antiquity, Sather Classical Lectures*. Berkeley, CA: University of California Press.

Shapere, Dudley. 1977. "Scientific Theories and their Domains." In *The Structure of Scientific Theories*, edited by Frederick Suppe, 518–570. Urbana, Chicago, London: University of Illinois Press.

Shubin, N., C. Tabin, and S. Carroll. 2009. "Deep Homology and the Origins of Evolutionary Novelty." *Nature* 457 (7231): 818–823.

Siddall, Mark E. 2001. "Philosophy and Phylogenetic Inference: A Comparison of Likelihood and Parsimony Methods in the Context of Karl Popper's Writings on Corroboration." *Cladistics* 17 (4): 395–399.

Simpson, George Gaylord. 1961. *Principles of Animal Taxonomy*. New York: Columbia University Press.

Simpson, P.K. 2002. "Fuzzy Min-max Neural Networks. I. Classification." *Neural Networks, IEEE Transactions on* 3 (5): 776–786.

Singh, Gurcharan. 2004. *Plant Systematics: An Integrated Approach*. Enfield, NH: Science Publishers.

Slaughter, Mary M. 1982. Universal Languages and Scientific Taxonomy in the Seventeenth Century. New York: Cambridge University Press.

Sloan, Phillip R. 1985. "From Logical Universals to Historical Individuals: Buffon's Idea of Biological Species." In *Histoire du Concept D'Espece dans les Sciences de la Vie*, 101–140. Paris: Fondation Singer-Polignac.

Small, Ernest. 1989. "Systematics of Biological Systematics (Or, Taxonomy of Taxonomy)." *Taxon* 38 (3): 335–356.

Sneath, P.H.A., and Robert R. Sokal. 1973. *Numerical Taxonomy: The Principles and Practice of Numerical Classification, A Series of books in biology*. San Francisco: W. H. Freeman.

Snyder, Laura J. 2006. *Reforming Philosophy: A Victorian Debate on Science and Society*. Chicago: University of Chicago Press.

Sober, Elliott. 2008. *Evidence and Evolution: The Logic Behind the Science*. Cambridge, UK: Cambridge University Press.

Sober, Elliott. 1999. "Modus Darwin." *Biology and Philosophy* 14 (2): 253–278.

Sober, Elliott. 1988. *Reconstructing the Past: Parsimony, Evolution, and Inference*. Cambridge, MA: MIT Press.

Sober, Elliott. 1980. "Evolution, Population Thinking, and Essentialism." *Philosophy of Science* 47: 350–383.

Soil Classification Working Group. 1998. *The Canadian System of Soil Classification*. 3rd edn. Vol. 1646, Agriculture and Agri-Food Canada Publication. Ottawa: NRC Research Press.

Soil Survey Staff. 1975. *Soil Taxonomy: A Basic System of Soil Classification for Making and Interpreting Soil Surveys, Issue 436 of Agriculture Handbook*. Washington DC: Soil conservation service U. S. Department of Agriculture.

Sokal, Robert R., and P.H.A. Sneath. 1963. *Principles of Numerical Taxonomy, A Series of Books in Biology*. San Francisco, CA: W. H. Freeman.

Spencer, Herbert. 1864. *The Classification of the Sciences*. New York: D. Appleton and company.

Sprague, T.A., and E. Nelmes. 1928/1931. "The Herbal of Leonhart Fuchs." *Journal of the Linnean Society. Botany* 48: 545–642.

Stafleu, Franz Antonie. 1971. *Linnaeus and the Linnaeans: The Spreading of their ideas in Systematic Botany, 1735–1789, Regnum vegetabile*, v. 79. Utrecht: Oosthoek.

Stamos, David N. 2003. *The Species Problem: Biological Species, Ontology, and the Metaphysics of Biology*. Lanham, MD: Lexington Books.

Stamos, David N. 2002. "Species, Languages, and the Horizontal/vertical Distinction." *Biology and Philosophy* 17: 171–198.

Steigerwald, J. 2002. "Goethe's Morphology: Urphänomene and Aesthetic Appraisal." *Journal of the History of Biology* 35: 291–238.

Sterelny, Kim, and Paul E. Griffiths. 1999. *Sex and Death: An Introduction to Philosophy of Biology, Science and its Conceptual Foundations*. Chicago, Ill.: University of Chicago Press.

Stevens, Peter F. 1994. *The Development of Biological Systematics: Antoine-Laurent de Jussieu, Nature, and the Natural System*. New York: Columbia University Press.

Stevens, Peter F. 1983. "Augustin Augier's 'Arbre Botanique' (1801), a Remarkable Early Botanical Representation of the Natural System." *Taxon* 32 (2): 203–211.

Swainson, William. 1835. *A Treatise on the Geography and Classification of Animals*. Edited by Rev. D. Lardner, The Cabinet Cyclopaedia: Natural History. London: Longman Rees, Orme, Brown and Longman.

Swainson, William. 1834. *Preliminary Discourse on the Study of Natural History*. London: Longman, Rees, Orme, Brown, Green and Longman.

Toon, Adam. 2012. *Models as Make-believe: Imagination, Fiction and Scientific Representation*. Edited by Steven French, New Directions in the Philosophy of Science. Basingstoke: Palgrave Macmillan.

Toulmin, Stephen. 1970. "Does the Distinction between Normal and Revolutionary Science hold Water?" In *Criticism and the Growth of Knowledge*, edited by I. Lakatos and I. Musgrave, 39–48. Cambridge, UK: Cambridge University Press.

Turner, G.F. 2002. "Parallel Speciation, Despeciation and Respeciation: Implications for Species Definition." *Fish and Fisheries* 3 (3): 225–229(0).

Turrill, Walter B. 1942. "Taxonomy and Phylogeny." *Botanical Review* 8 (4): 247–270.

Turrill, Walter B. 1940. "Experimental and Synthetic Plant Taxonomy." In *The New Systematics*, edited by Julian Huxley, 47–72. London: Oxford University Press.

Turrill, Walter B. 1935. "The Investigation of Plant Species." *Proceedings of the Linnean Society of London* 147: 104–105.

Tversky, Amos, and Itamar Gati. 1978. "Studies of Similarity." In *Cognition and Categorization*, edited by Eleanor Rosch and Barbara B. Lloyd, 79–98. Hillsdale, NJ: Lawrence Erlbaum Associates.

van Fraassen, Bas C. 2008. *Scientific Representation: Paradoxes of Perspective*. Oxford: Clarendon Press.

van Fraassen, Bas C. 1980. *The Scientific Image*. Oxford: Clarendon Press.

van Gelder, T. 1998a. "Disentangling Dynamics, Computation, and Cognition – Response." *Behavioral and Brain Sciences* 21 (5): 654–665.

van Gelder, T. 1998b. "The Dynamical Hypothesis in Cognitive Science." *Behavioral and Brain Sciences* 21 (5): 616–665.

Varfolomeev, L. 2010. "Still in our Memory: On the 150th Anniversary of the Birth and the 110th Anniversary of the Death of Nikolai Mikhailovich Sibirtsev." *Eurasian Soil Science* 43 (11): 1294–1300.

Venn, John. 1866. *The Logic of Chance: An Essay on the Foundations and Province of the Theory of Probability, with Especial Reference to Its Application to Moral and Social Science*. London: Macmillan.

Wagner, G. P. 2007. "The Developmental Genetics of Homology." *Nature Reviews Genetics* 8 (6): 473–479.

Wagner, G. P. 1989. "The Biological Homology Concept." *Annual Review of Ecology and Systematics* 20 (1): 51–69.

Wagner, Günter P., and Peter F. Stadler. 2003. "Quasi-Independence, Homology and the Unity of Type: A Topological Theory of Characters." *Journal of Theoretical Biology* 220 (4): 505–527.

Walsh, Denis. 2006. "Evolutionary Essentialism." *British Journal for the Philosophy of Science* 57 (2): 425–448.

Watt, Alex S. 1947. "Pattern and Process in the Plant Community." *Journal of Ecology* 35 (1/2): 1–22.

Weisberg, Michael. 2007. "Who is a Modeler?" *British Journal for the History of Philosophy* 58: 207–233.

Whately, Richard. 1875. *Elements of Logic*. 9th (octavo) edn. London: Longmans, Green and Co. Original edition, 1826.

Wheeler, Quentin D. 2005. "Losing the Plot: DNA 'Barcodes' and Taxonomy." *Cladistics* 21 (4): 405–407.

Whewell, William. 1840. *The Philosophy of the Inductive Sciences: Founded Upon Their History*. 2 vols. London: John W. Parker.

Whewell, William. 1837. *History of the Inductive Sciences*. London: Parker.

Whitehead, Alfred North. 1938. *Science and the Modern World*. Pelican ed. Harmondsworth: Penguin.

Whitehead, Alfred North, and Bertrand Russell. 1910. *Principia Mathematica*. Cambridge: Cambridge University Press.

Wilkerson, T.E. 1993. "Species, Essences and the Names of Natural Kinds." *Philosophical Quarterly*

Wilkins, John. 1668. *An essay towards a real character, and a philosophical language. By John Wilkins*. London: printed for Sa: Gellibrand, and for John Martyn printer to the Royal Society.

Wilkins, John S. forthcoming. "Biological Essentialism and the Tidal Change of Natural Kinds." *Science and Education* 22 (2): 221–240.

Wilkins, John S. 2013. "Essentialism in Biology." In *Philosophy of Biology: A Companion for Educators*, edited by Kostas Kampourakis, 395–419. Dordrecht: Springer.

Wilkins, John S. 2011. "Philosophically Speaking, How Many Species Concepts are There?" *Zootaxa* 2765: 58–60.

Wilkins, John S. 2010. "What is a species? Essences and Generation." *Theory in Biosciences* 129: 141–148.

Wilkins, John S. 2009a. *Defining Species: A Sourcebook from Antiquity to Today*, American University Studies. V, Philosophy. New York: Peter Lang.

Wilkins, John S. 2009b. *Species: A History of the Idea, Species and Systematics*. Berkeley: University of California Press.

Wilkins, John S. 2008. "The Adaptive Landscape of Science." *Biology and Philosophy* 23 (5): 659–671.

Wilkins, John S. 2007a. "The Concept and Causes of Microbial Species." *Studies in History and Philosophy of the Life Sciences* 28 (3): 389–408.

Wilkins, John S. 2007b. "The Dimensions, Modes and Definitions of Species and Speciation." *Biology and Philosophy* 22 (2): 247–266.

Wilkins, John S. 2003. "How to be a Chaste Species Pluralist-realist: The Origins of Species Modes and the Synapomorphic Species Concept." *Biology and Philosophy* 18: 621–638.

Wilkins, John S. 2002. "Darwinism as Metaphor and Analogy: Language as a Selection Process." *Selection: Molecules, Genes, Memes* 3 (1): 57–74.

Wilkins, John S., and Paul E. Griffiths. 2013. "Evolutionary Debunking Arguments in Three Domains: Fact, Value, and Religion." In *A New Science of Religion*, edited by J. Maclaurin and G. Dawes, 133–146. Chicago: University of Chicago Press.

Will, Kipling W., Brent D. Mishler, and Quentin D. Wheeler. 2005. "The Perils of DNA Barcoding and the Need for Integrative Taxonomy." *Systematic Biology* 54 (5): 844.

Will, Kipling W., and Daniel Rubinoff. 2004. "Myth of the Molecule: DNA Barcodes for Species cannot Replace Morphology for Identification and Classification." *Cladistics* 20 (1): 47–55.

Williams, David M., Malte C. Ebach, and Gareth Nelson. 2008. *Foundations of Systematics and Biogeography*. New York, N.Y.: Springer.

Williams, David M. 2004. "Homology and Homologues, Cladistics and Phenetics: 150 Years of Progress." In *Milestones in Systematics*, edited by David M. Williams and Peter L. Forey, 191–224. Boca Raton, London: Taylor and Francis/CRC Press.

Williams, David M., and Malte C. Ebach. in press. "Patterson's Curse, Molecular Homology, and the Data Matrix." In *The Evolution of Phylogenetic Systematics*, edited by Andrew Hamilton. Berkeley CA: University of California Press.

Williams, David M., and Malte C. Ebach. 2012. "Confusing Homologs as Homologies: A Reply to 'On homology'." *Cladistics* 28 (3): 223–224.

Wilson, Edward O. 1995. *Naturalist*. London: Allen Lane The Penguin Press.

Wilson, Robert A. 1999a. "Realism, Essence, and Kind: Resuscitating Species Essentialism?" In *Species, New Interdisciplinary Essays*, edited by R. A. Wilson, 187–208. Cambridge, MA: MIT Press.

Wilson, Robert A. 1999b. *Species: New Interdisciplinary Essays*. Cambridge, MA: MIT Press.

Winsor, Mary Pickard 2009. "Taxonomy was the Foundation of Darwin's Evolution." *Taxon* 58 (1): 43–49.

Winsor, Mary Pickard. 2006a. "The Creation of the Essentialism Story: An Exercise in Metahistory." *History and Philosophy of the Life Sciences* . 28: 149–174.

Winsor, Mary Pickard. 2006b. "Linnaeus' Biology was not Essentialist." *Annals of the Missouri Botanical Garden* 93 (1): 2–7.

Winsor, Mary Pickard. 2004 . "Setting up Milestones: Sneath on Adanson and Mayr on Darwin." In *Milestones in Systematics: Essays from a Symposium held within the 3rd Systematics Association Biennial Meeting*, September 2001, edited by David M. Williams and Peter L. Forey, 1–17. London: Systematics Association.

Winsor, Mary Pickard. 2003. "Non-essentialist Methods in Pre-Darwinian Taxonomy." *Biology & Philosophy* 18: 387–400.

Winsor, Mary Pickard 2000. "Species, Demes, and the Omega Taxonomy: Gilmour and The New Systematics." *Biology and Philosophy* 15 (3): 349–388.

Winston, Judith E. 1999. *Describing Species: Practical Taxonomic Procedures for Biologists*. New York: Columbia University Press.

Witmer, Lawrence M. 1995. "The Extant Phylogenetic Bracket and the Importance of Reconstructing Soft Tissues in Fossils." In *Functional Morphology in Vertebrate Paleontology*, edited by Jeff Thomason, 19–33. New York: University of Cambridge Press.

Wittgenstein, Ludwig. 1968. *Philosophical Investigations*. Translated by G.E.M. Anscombe. Repr. of [3rd ed.] English text, with index edn. Oxford: Basil Blackwell.

Wittgenstein, Ludwig. 1922. *Tractatus Logico-philosophicus*. London: Kegan Paul, Trench, Trubner and Co.

Wolpert, David H., and William G. Macready. 1997. "No Free Lunch Theorems for Search." *IEEE Transactions on Evolutionary Computation* 1 (1): 67–82.

Wolpert, David H., and William G. Macready. 1995. *No Free Lunch Theorems for Search*. Sante Fe, NM: Santa Fe Institute.

Woodger, Joseph Henry. 1937. *The Axiomatic Method in Biology*. Cambridge UK: Cambridge University Press.

World Meteorological Organization. 1975. *International Cloud Atlas*. Secretariat of the World Meteorological Organization.

World Meteorological Organization. 1956. *International Cloud Atlas*. Vol. 1. Geneva: World Meteorological Organization, Atar S.A.

Yoon, Carol Kaesuk. 2010. *Naming Nature: The Clash between Instinct and Science*. New York: W. W. Norton.

Zalta, Edward N. 1988. *Abstract Objects: An Introduction to Axiomatic Metaphysics*. Dordrecht: D. Reidel.

Zangerl, Rainer. 1948. "The Methods of Comparative Anatomy and its Contribution to the Study of Evolution." *Evolution* 2 (4): 351–374.

Index

Printed in the United States
by Baker & Taylor Publisher Services